SCHAUM'S *Easy* OUTLINES

APPLIED PHYSICS

Other Books in Schaum's Easy Outlines Series Include:

SCHAUM'S *Easy* OUTLINES

APPLIED PHYSICS

BASED ON SCHAUM'S
Outline of Theory and Problems of
Applied Physics (Third Edition)
BY ARTHUR BEISER, Ph.D.

ABRIDGEMENT EDITOR
GEORGE J. HADEMENOS, Ph.D.

SCHAUM'S OUTLINE SERIES
McGRAW-HILL

New York   Chicago   San Francisco   Lisbon   London   Madrid
Mexico City   Milan   New Delhi   San Juan
Seoul   Singapore   Sydney   Toronto

ARTHUR BEISER received his Ph.D. from New York University, where he subsequently served as Assistant and Associate Professor of Physics. He has been a consultant to various industrial firms and government agencies and is the author of more than a dozen textbooks of physics and mathematics.

GEORGE J. HADEMENOS has taught at the University of Dallas and done research at the University of Massachusetts Medical Center and the University of California at Los Angeles. He holds a B.S. degree from Angelo State University and both M.S. and Ph.D degrees from the University of Texas at Dallas. He is the author of several books in the *Schaum's Outline* and *Schaum's Easy Outline* series.

1 2 3 4 5 6 7 8 9   DOC/DOC   0 9 8 7 6 5 4 3

ISBN 0-07-138978-3

Contents

Chapter 1
VECTORS

Scalar and Vector Quantities

A *scalar quantity* has only magnitude and is completely specified by a number and a unit. Examples are mass (a stone has a mass of 2 kg), volume (a bottle has a volume of 1.5 liters), and frequency (house current has a frequency of 60 Hz). Scalar quantities of the same kind are added by using ordinary arithmetic.

A *vector quantity* has both magnitude and direction. Examples are displacement (an airplane has flown 200 km to the southwest), velocity (a car is moving 60 km/h to the north), and force (a person

1

applies an upward force of 25 newtons to a package). Symbols of vector quantities are printed in boldface type (**v** = velocity, **F** = force). When vector quantities of the same kind are added, their directions must be taken into account.

Vector Addition: Graphical Method

A *vector* is represented by an arrow whose length is proportional to a certain vector quantity and whose direction indicates the direction of the quantity.

To add vector **B** to vector **A**, draw **B** so that its tail is at the head of **A**. The vector sum **A** + **B** is the vector **R** that joins the tail of **A** and the head of **B** (Figure 1-1). Usually, **R** is called the *resultant* of **A** and **B**. The order in which **A** and **B** are added is not significant, so that **A** + **B** = **B** + **A** (Figures 1-1 and 1-2).

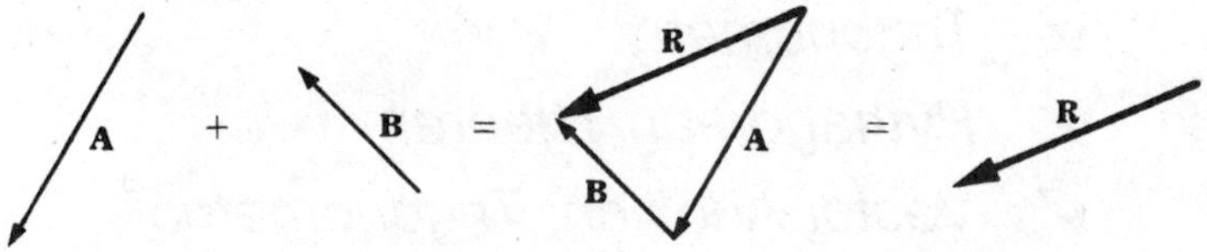

Figure 1-1

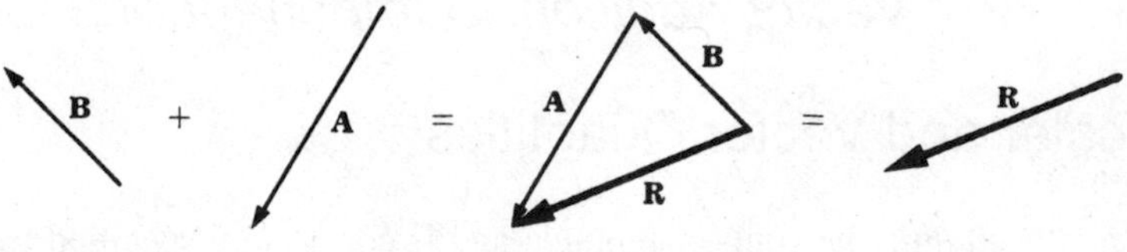

Figure 1-2

Exactly the same procedure is followed when more than two vectors of the same kind are to be added. The vectors are strung together head to tail (being careful to preserve their correct lengths and directions), and the resultant **R** is the vector drawn from the tail of the first vector to the head of the last. The order in which the vectors are added does not matter (Figure 1-3).

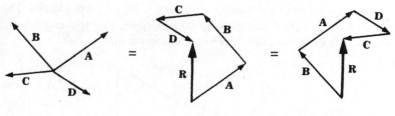

Figure 1-3

Solved Problem 1.1 A woman walks eastward for 5 km and then north-ward for 10 km. How far is she from her starting point? If she had walked directly to her destination, in what direction would she have headed?

Solution. From Figure 1-4, the length of the resultant vector **R** corre-sponds to a distance of 11.2 km, and a protractor shows that its direction is 27° east of north.

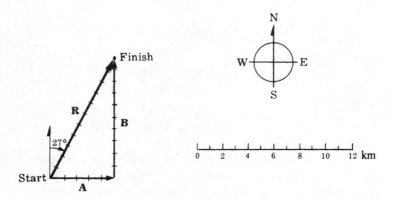

Figure 1-4

Trigonometry

Although it is possible to determine the magnitude and direction of the resultant of two or more vectors of the same kind graphically with ruler or protractor, this procedure is not very exact. For accurate results, it is necessary to use trigonometry.

A *right triangle* is a triangle whose two sides are perpendicular. The *hypotenuse* of a right triangle is the side opposite the right angle, as in Figure 1-5; the hypotenuse is always the longest side.

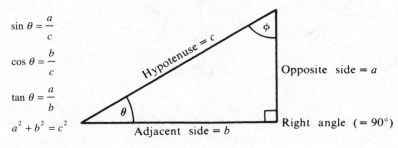

$$\sin \theta = \frac{a}{c}$$

$$\cos \theta = \frac{b}{c}$$

$$\tan \theta = \frac{a}{b}$$

$$a^2 + b^2 = c^2$$

Hypotenuse $= c$

Opposite side $= a$

Adjacent side $= b$

Right angle $(= 90°)$

Figure 1-5

The three basic trigonometric functions—the sine, cosine, and tangent of an angle—are defined in terms of the right triangle of Figure 1-5 as follows:

$$\sin \theta = \frac{a}{c} = \frac{\text{opposite side}}{\text{hypotenuse}}$$

$$\cos \theta = \frac{b}{c} = \frac{\text{adjacent side}}{\text{hypotenuse}}$$

$$\tan \theta = \frac{a}{b} = \frac{\text{opposite side}}{\text{adjacent side}} = \frac{\sin \theta}{\cos \theta}$$

The *inverse* of a trigonometric function is the angle whose function is given. Thus the inverse of $\sin \theta$ is the angle θ. The names and abbreviations of the inverse trigonometric functions are as follows:

$$\sin \theta = x$$

$$\theta = \arcsin x = \sin^{-1} x = \text{angle whose sine is } x$$

$$\cos \theta = y$$

$$\theta = \arccos y = \cos^{-1} y = \text{angle whose cosine is } y$$

$$\tan \theta = z$$

$$\theta = \arctan z = \tan^{-1} z = \text{angle whose tangent is } z$$

Remember

In trigonometry, an expression such as $\sin^{-1}x$ does *not* signify $1/(\sin x)$, even though in algebra, the exponent -1 signifies a reciprocal.

Pythagorean Theorem

The *Pythagorean theorem* states that the sum of the squares of the short sides of a right triangle is equal to the square of its hypotenuse. For the triangle of Figure 1-5,

$$a^2 + b^2 = c^2$$

Hence, we can always express the length of any of the sides of a right triangle in terms of the lengths of the other sides:

$$a = \sqrt{c^2 - b^2} \quad b = \sqrt{c^2 - a^2} \quad c = \sqrt{a^2 + b^2}$$

Another useful relationship is that the sum of the interior angles of any triangle is 180°. Since one of the angles in a right triangle is 90°, the sum of the other two must be 90°. Thus, in Figure 1-5, $\theta + \phi = 90°$.

Of the six quantities that characterize a triangle—three sides and three angles—we must know the values of at least three, including one of the sides, in order to calculate the others. In a right triangle, one of the angles is always 90°, so all we need are the lengths of any two sides or the length of one side plus the value of one of the other angles to find the other sides and angles.

Solved Problem 1.2 Find the values of the sine, cosine, and tangent of angle θ in Figure 1-6.

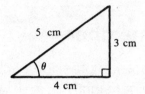

Figure 1-6

Solution.

$$\sin\ \theta = \frac{\text{opposite side}}{\text{hypotenuse}} = \frac{3\ \text{cm}}{5\ \text{cm}} = 0.6$$

$$\cos\ \theta = \frac{\text{adjacent side}}{\text{hypotenuse}} = \frac{4\ \text{cm}}{5\ \text{cm}} = 0.8$$

$$\tan\ \theta = \frac{\text{opposite side}}{\text{adjacent side}} = \frac{3\ \text{cm}}{4\ \text{cm}} = 0.75$$

Vector Addition: Trigonometric Method

It is easy to apply trigonometry to find the resultant **R** of two vectors **A** and **B** that are perpendicular to each other. The magnitude of the resultant is given by the Pythagorean theorem as:

$$R = \sqrt{A^2 + B^2}$$

and the angle between **R** and **A** (Figure 1-7) may be found from

$$\tan\ \theta = \frac{B}{A}$$

by examining a table of tangents or by using a calculator to determine $\tan^{-1}\frac{B}{A}$.

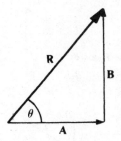

Figure 1-7

Resolving a Vector

Just as two or more vectors can be added to yield a single resultant vector, so it is possible to break up a single vector into two or more other vectors. If vectors **A** and **B** are together equivalent to vector **C**, then vector **C** is equivalent to the two vectors **A** and **B** (Figure 1-8). When a vector is replaced by two or more others, the process is called *resolving* the vector, and the new vectors are known as the *components* of the initial vector.

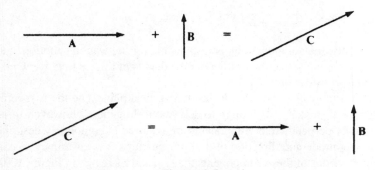

Figure 1-8

The components into which a vector is resolved are nearly always chosen to be perpendicular to one another. Figure 1-9 shows a wagon being pulled by a man with force **F**. Because the wagon moves horizontally, the entire force is not effective in influencing its motion.

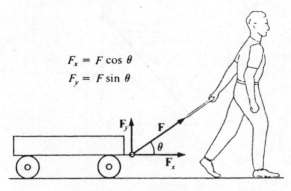

$$F_x = F \cos \theta$$
$$F_y = F \sin \theta$$

Figure 1-9

The force **F** may be resolved into two component vectors $\mathbf{F}_x$ and $\mathbf{F}_y$, where

$\mathbf{F}_x$ = horizontal component of **F**

$\mathbf{F}_y$ = vertical component of **F**

The magnitudes of these components are

$$F_x = F \cos \theta \qquad F_y = F \sin \theta$$

Evidently, the component $\mathbf{F}_x$ is responsible for the wagon's motion, and if we were interested in working out the details of this motion, we would need to consider only $\mathbf{F}_x$.

In Figure 1-9, the force **F** lies in a vertical plane, and the two components $\mathbf{F}_x$ and $\mathbf{F}_y$ are enough to describe it. In general, however, three mutually perpendicular components are required to completely describe the magnitude and direction of a vector quantity. It is customary to call the directions of these components the x, y, and z axes, as in Figure 1-10. The component of some vector **A** in these directions are accordingly denoted $\mathbf{A}_x$, $\mathbf{A}_y$, and $\mathbf{A}_z$. If a component falls on the negative part of an axis, its magnitude is considered negative. Thus, if $\mathbf{A}_z$ were downward in Figure 1-10 instead of upward and its length were equivalent to, say, 12 N, we would write $A_z = -12$ N. (The newton (N) is the SI unit of force; it is equal to 0.225 lb.)

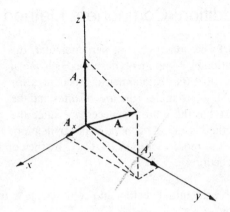

Figure 1-10

Solved Problem 1.3 The man in Figure 1-9 exerts a force of 100 N on the wagon at an angle of $\theta = 30°$ above the horizontal. Find the horizontal and vertical components of this force.

Solution. The magnitudes of $\mathbf{F}_x$ and $\mathbf{F}_y$ are, respectively,

$$F_x = F\cos\theta = (100 \text{ N})(\cos\ 30°) = 86.6 \text{ N}$$
$$F_y = F\sin\theta = (100 \text{ N})(\sin\ 30°) = 50.0 \text{ N}$$

We note that $F_x + F_y = 136.6$ N although $\mathbf{F}$ itself has the magnitude $F = 100$ N. What is wrong? The answer is that nothing is wrong; because F_x and F_y are just the *magnitudes* of the vectors $\mathbf{F}_x$ and $\mathbf{F}_y$, it is meaningless to add them. However, we can certainly add the *vectors* $\mathbf{F}_x$ and $\mathbf{F}_y$ to find the magnitude of their resultant $\mathbf{F}$. Because $\mathbf{F}_x$ and $\mathbf{F}_y$ are perpendicular,

$$F = \sqrt{F_x^2 + F_y^2} = \sqrt{(86.6 \text{ N})^2 + (50.0 \text{ N})^2} = 100 \text{ N}$$

as we expect.

Vector Addition: Component Method

When vectors to be added are not perpendicular, the method of addition by components described below can be used. There do exist trigonometric procedures for dealing with oblique triangles (the *law of sines* and the *law of cosines*), but these are not necessary since the component method is entirely general in its application.

To add two or more vectors **A**, **B**, **C**, ... by the component method, follow this procedure:

1. Resolve the initial vectors into components in the x, y, and z directions.
2. Add the components in the x direction to give $\mathbf{R}_x$, add the components in the y direction to give $\mathbf{R}_y$, and add the components in the z direction to give $\mathbf{R}_z$. That is, the magnitudes of $\mathbf{R}_x$, $\mathbf{R}_y$, and $\mathbf{R}_z$ are given by, respectively,

$$R_x = A_x + B_x + C_x + \cdots$$
$$R_y = A_y + B_y + C_y + \cdots$$
$$R_z = A_z + B_z + C_z + \cdots$$

3. Calculate the magnitude and direction of the resultant **R** from its components $\mathbf{R}_x$, $\mathbf{R}_y$, and $\mathbf{R}_z$ by using the Pythagorean theorem:

$$R = \sqrt{R_x^2 + R_y^2 + R_z^2}$$

If the vectors being added all lie in the same plane, only two components need to be considered.

Chapter 2
MOTION

Velocity

The *velocity* of a body is a vector quantity that describes both how fast it is moving and the direction in which it is headed.

In the case of a body traveling in a straight line, its velocity is simply the rate at which it covers distance. The *average velocity* v̄ of such a body when it covers the distance s in the time t is

$$\bar{v} = \frac{s}{t}$$

$$\text{Average velocity} = \frac{\text{distance}}{\text{time}}$$

The average velocity of a body during the time t does not completely describe its motion, however, because during the time t, it may some-

11

times have gone faster than $\bar{v}$ and sometimes slower. The velocity of a body at any given moment is called its *instantaneous velocity* and is given by

$$v_{inst} = \frac{\Delta s}{\Delta t}$$

Here, Δs is the distance the body has gone in the very short time interval Δt at the specified moment. (Δ is the capital Greek letter *delta*.) Instantaneous velocity is what a car's speedometer indicates.

When the instantaneous velocity of a body does not change, it is moving at *constant velocity*. For the case of constant velocity, the basic formula is

$$s = vt$$

Distance = (constant velocity)(time)

Solved Problem 2.1 The velocity of sound in air at sea level is about 343 m/s. If a person hears a clap of thunder 3.00 s after seeing a lightning flash, how far away was the lightning?

Solution. The velocity of light is so great compared with the velocity of sound that the time needed for the light of the flash to reach the person can be neglected. Hence

$$s = vt = (343 \text{ m/s})(3.00 \text{ s}) = 1029 \text{ m} = 1.03 \text{ km}$$

Acceleration

A body whose velocity is changing is accelerated. A body is accelerated when its velocity is increasing, decreasing, or changing its direction.

The *acceleration* of a body is the rate at which its velocity is changing. If a body moving in a straight line has a velocity of v_0 at the start of a certain time interval t and of v at the end, its acceleration is

$$a = \frac{v - v_0}{t}$$

$$\text{Acceleration} = \frac{\text{velocity change}}{\text{time}}$$

A positive acceleration means an increase in velocity; a negative acceleration (sometimes called *deceleration*) means a decrease in velocity. Only constant accelerations are considered here.

The defining formula for acceleration can be rewritten to give the final velocity v of an accelerated body:

$$v = v_0 + at$$

Final velocity = initial velocity + (acceleration)(time)

We can also solve for the time t in terms of v_0, v, and a:

$$t = \frac{v - v_0}{a}$$

$$\text{Time} = \frac{\text{velocity change}}{\text{acceleration}}$$

Velocity has the dimensions of distance/time. Acceleration has the dimensions of velocity/time or distance/time^2. A typical acceleration unit is the meter/second^2 (meter per second squared). Sometimes two different time units are convenient; for instance, the acceleration of a car that goes from rest to 90 km/h in 10 s might be expressed as $a = 9$ (km/h)/s.

Solved Problem 2.2 A car starts from rest and reaches a final velocity of 40 m/s in 10 s. (*a*) What is its acceleration? (*b*) If its acceleration remains the same, what will its velocity be 5 s later?

Solution. (*a*) Here $v_0 = 0$. Hence

$$a = \frac{v}{t} = \frac{40 \text{ m/s}}{10 \text{ s}} = 4 \text{ m/s}^2$$

(*b*) Now $v_0 = 40$ m/s, so

$$v = v_0 + at = 40 \text{ m/s} + (4 \text{ m/s}^2)(5 \text{ s}) = 40 \text{ m/s} + 20 \text{ m/s} = 60 \text{ m/s}$$

Distance, Velocity, and Acceleration

Let us consider a body whose velocity is v_0 when it starts to be accelerated at a constant rate. After time t, the final velocity of the body will be

$$v = v_0 + at$$

How far does the body go during the time interval t? The average velocity $\bar{v}$ of the body is

$$\bar{v} = \frac{v_0 + v}{2}$$

and so

$$s = \bar{v}t = \left(\frac{v_0 + v}{2}\right)t$$

Since $v = v_0 + at$, another way to specify the distance covered during t is

$$s = \left(\frac{v_0 + v_0 + at}{2}\right)t = v_0 t + \frac{1}{2}at^2$$

If the body is accelerated from rest, $v_0 = 0$ and

$$s = \frac{1}{2}at^2$$

Another useful formula gives the final velocity of a body in terms of its initial velocity, its acceleration, and the distance it has traveled during the acceleration:

$$v^2 = v_0^2 + 2as$$

This can be solved for the distance s to give

$$s = \frac{v^2 - v_0^2}{2a}$$

In the case of a body that starts from rest, $v_0 = 0$ and

$$v = \sqrt{2as} \qquad s = \frac{v^2}{2a}$$

Table 2.1 summarizes the formulas for motion under constant acceleration.

Distance	Final Velocity
$s = \left(\dfrac{v_0 + v}{2}\right)t$	$v = v_0 + at$
$s = v_0 t + \frac{1}{2}at^2$	$v^2 = v_0^2 + 2as$

Table 2.1

Acceleration of Gravity

All bodies in free fall near the earth's surface have the same downward acceleration of

$$g = 9.8 \text{ m/s}^2 = 32 \text{ ft/s}^2$$

A body falling from rest in a vacuum thus has a velocity of 32 ft/s at the end of the first second, 64 ft/s at the end of the next second, and so forth. The farther the body falls, the faster it moves.

You Need to Know

A body in free fall has the same downward acceleration whether it starts from rest or has an initial velocity in some direction.

The presence of air affects the motion of falling bodies partly through buoyancy and partly through air resistance. Thus two different objects falling in air from the same height will not, in general, reach the ground at exactly the same time. Because air resistance

increases with velocity, eventually a falling body reaches a *terminal velocity* that depends on its mass, size, and shape, and it cannot fall any faster than that.

Falling Bodies

When buoyancy and air resistance can be neglected, a falling body has the constant acceleration g and the formulas for uniformly accelerated motion apply. Thus a body dropped from rest has the velocity

$$v = gt$$

after time t, and it has fallen through a vertical distance of

$$h = \frac{1}{2}gt^2$$

From the latter formula, we see that

$$t = \sqrt{\frac{2h}{g}}$$

and so the velocity of the body is related to the distance it has fallen by $v = gt$, or

$$v = \sqrt{2gh}$$

To reach a certain height h, a body thrown upward must have the same initial velocity as the final velocity of a body falling from that height, namely, $v = \sqrt{2gh}$.

Solved Problem 2.3 What velocity must a ball have when thrown upward if it is to reach a height of 15 m?

Solution. The upward velocity the ball must have is the same as the downward velocity the ball would have if dropped from that height. Hence

$$v = \sqrt{2gh} = \sqrt{(2)(9.8 \text{ m}/\text{s}^2)(15 \text{ m})} = \sqrt{294 \text{ m}^2/s^2} = 17 \text{ m}/\text{s}$$

Projectile Motion

The formulas for straight-line motion can be used to analyze the horizontal and vertical aspects of a projectile's flight separately because these are independent of each other. If air resistance is neglected, the horizontal velocity component v_x remains constant during the flight. The effect of gravity on the vertical component v_y is to provide a downward acceleration. If v_y is initially upward, v_y first decreases to 0 and then increases in the downward direction.

The range of a projectile launched at an angle θ above the horizontal with initial velocity v_0 is

$$R = \frac{v_0^2}{g}\sin 2\theta$$

The time of flight is

$$T = \frac{2v_0 \sin \theta}{g}$$

If θ_1 is an angle other than 45° that corresponds to a range R, then a second angle θ_2 for the same range is given by

$$\theta_2 = 90° - \theta_1$$

as shown in Figure 2–1.

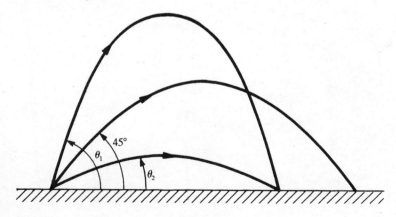

Figure 2-1

Solved Problem 2.4 A football is thrown with a velocity of 10 m/s at an angle of 30° above the horizontal. (a) How far away should its intended receiver be? (b) What will the time of flight be?

Solution.
(a)

$$R = \frac{v_0^2}{g} \sin 2\theta = \left[\frac{(10 \text{ m/s})^2}{9.8 \text{ m/s}^2}\right](\sin 60°) = 8.8 \text{ m}$$

(b)

$$T = \frac{2v_0 \sin \theta}{g} = \frac{(2)(10 \text{ m/s})(\sin 30°)s}{9.8 \text{ m/s}^2} = 1.02 \text{ s}$$

Chapter 3
NEWTON'S LAWS OF MOTION

IN THIS CHAPTER:

- ✔ *First Law of Motion*
- ✔ *Mass*
- ✔ *Second Law of Motion*
- ✔ *Weight*
- ✔ *British System of Units*
- ✔ *Free-Body Diagrams and Tension*
- ✔ *Third Law of Motion*
- ✔ *Static and Kinetic Friction*
- ✔ *Coefficient of Friction*

First Law of Motion

According to Newton's *first law of motion*, if no net force acts on it, a body at rest remains at rest and a body in motion remains in motion at constant velocity (that is, at constant speed in a straight line).

This law provides a definition of *force*: A force is any influence that can change the velocity of a body.

Two or more forces act on a body without affecting its velocity if the forces cancel one another out.

19

What is needed for a velocity change is a *net force*, or *unbalanced force*. To accelerate something, a net force must be applied to it. Conversely, every acceleration is due to the action of a net force.

Mass

The property a body has of resisting any change in its state of rest or uniform motion is called *inertia*. The inertia of a body is related to what we think of as the amount of matter it contains. A quantitative measure of inertia is *mass*: The more mass a body has, the less its acceleration when a given net force acts on it. The SI unit of mass is the *kilogram* (kg).

 **Note!**

A liter of water, which is 1.057 quarts, has a mass of almost exactly 1 kg.

Second Law of Motion

According to Newton's *second law of motion*, the net force acting on a body equals the product of the mass and the acceleration of the body. The direction of the force is the same as that of the acceleration.

In equation form,

$$\mathbf{F} = m\mathbf{a}$$

Net force is sometimes designated $\Sigma \mathbf{F}$, where Σ (Greek capital letter *sigma*) means "sum of." The second law of motion is the key to understanding the behavior of moving bodies since it links cause (force) and effect (acceleration) in a definite way.

In the SI system, the unit for force is the *newton* (N): A newton is that net force which, when applied to a 1-kg mass, gives it an acceleration of 1 m/s^2.

Solved Problem 3.1 A 10-kg body has an acceleration of 5 m/s^2. What is the net force acting on it?

Solution.

$$F = ma = (10 \text{ kg})(5 \text{ m/s}^2) = 50 \text{ N}$$

Weight

The *weight* of a body is the gravitational force with which the earth attracts the body. If a person weighs 600 N (135 lb), this means the earth pulls that person down with a force of 600 N. Weight (a vector quantity) is different from mass (a scalar quantity), which is a measure of the response of a body to an applied force. The weight of a body varies with its location near the earth (or other astronomical body), whereas its mass is the same everywhere in the universe.

The weight of a body is the force that causes it to be accelerated downward with the acceleration of gravity g. Hence, from the second law of motion, with $F = w$ and $a = g$,

$$w = mg$$
Weight = (mass)(acceleration of gravity)

Because g is constant near the earth's surface, the weight of a body there is proportional to its mass—a large mass is heavier than a small one.

British System of Units

In the British system, the unit of mass is the *slug* and the unit of force is the *pound* (lb). A net force of 1 lb acting on a mass of 1 slug gives it an acceleration of 1 ft/s^2. Table 3.1 shows how units of mass and force in the SI and British systems are related.

System of Units	To find mass m given weight w	To find weight w given mass m
SI	$m \text{ kg} = \dfrac{w \text{ N}}{9.8 \text{ m/s}^2}$	$w \text{ N} = (m \text{ kg})(9.8 \text{ m/s}^2)$
British	$m \text{ slugs} = \dfrac{w \text{ lb}}{32 \text{ ft/s}^2}$	$w \text{ lb} = (m \text{ slugs})(32 \text{ ft/s}^2)$

Table 3.1

Free-Body Diagrams and Tension

In all but the simplest problems that involve the second law of motion, it is helpful to draw a *free-body diagram* of the situation. This is a vector diagram that shows all of the forces that act *on* the body whose motion is being studied. Forces that the body exerts on anything else should not be included, since such forces do not affect the body's motion.

Forces are often transmitted by *cables*, a general term that includes strings, ropes, and chains. Cables can change the direction of a force with the help of a pulley while leaving the magnitude of the force unchanged. The *tension T* in a cable is the magnitude of the force that any part of the cable exerts on the adjoining part (Figure 3-1). The tension is the same in both directions in the cable, and *T* is the same along the entire cable if the cable's mass is small. Only cables of negligible mass will be considered here, so *T* can be thought of as the magnitude of the force that either end of a cable exerts on whatever it is attached to.

Solved Problem 3.2 Figure 3-2 shows a 5-kg block *A* which hangs from a string that passes over a frictionless pulley and is joined at its other end to a 12-kg block *B* that lies on a frictionless table.

(*a*) Find the acceleration of the two blocks. (*b*) Find the tension in the string.

Solution. (*a*) See Figure 3-2. The blocks have accelerations of the same magnitude *a* because they are joined by the string. The net force F_B on *B*

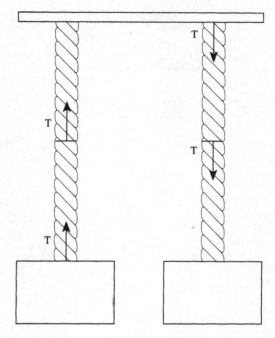

Figure 3-1

equals the tension T in the string. From the second law of motion, taking the left as the + direction so that a will come out positive,

$$F_B = T = m_B a$$

The net force F_A on A is the difference between its weight $m_A g$, which acts downward, and the tension T in the string, which acts upward on it. Taking downward as + so that the two accelerations have the same sign,

$$F_A = m_A g - T = m_A a$$

We now have two equations in the two unknowns, a and T. The easiest way to solve them is to start by substituting $T = m_B a$ from the first equation into the second. This gives

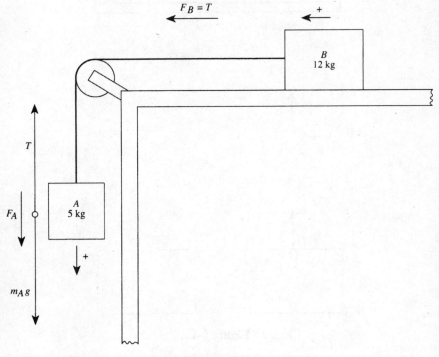

Figure 3-2

$$m_A g - T = m_A g - m_B a = m_A a$$

$$m_A g = (m_A + m_B)a$$

$$a = \frac{m_A g}{m_A + m_B} = \frac{(5 \text{ kg})(9.8 \text{ m}/\text{s}^2)}{5 \text{ kg} + 12 \text{ kg}} = 2.9 \text{ m}/\text{s}^2$$

(b) We can use either of the original equations to find the tension T. From the first,

$$T = m_B a = (12 \text{ kg})(2.9 \text{ m}/\text{s}^2) = 35 \text{ N}$$

Third Law of Motion

According to Newton's *third law of motion*, when one body exerts a force on another body, the second body exerts on the first an equal force in the opposite direction. The third law of motion applies to two different forces on two different bodies: the *action force* one body exerts on the other, and the equal but opposite *reaction force* the second body exerts on the first. Action and reaction forces never cancel each other out because they act on different bodies.

Solved Problem 3.3 A book rests on a table. (*a*) Show the forces acting on the table and the corresponding reaction forces. (*b*) Why do the forces acting on the table not cause it to move?

Solution.
(*a*) See Figure 3-3.

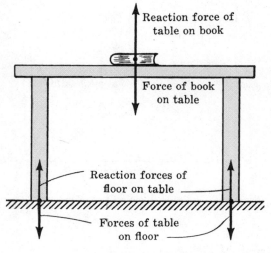

Figure 3-3

(*b*) The forces that act on the table have a vector sum of zero, so there is no net force acting on it.

Static and Kinetic Friction

Frictional forces act to oppose relative motion between surfaces that are in contact. Such forces act parallel to the surfaces.

Static friction occurs between surfaces at rest relative to each other. When an increasing force is applied to a book resting on a table, for instance, the force of static friction at first increases as well to prevent motion. In a given situation, static friction has a certain maximum value called *starting friction*. When the force applied to the book is greater than the starting friction, the book begins to move across the table. The *kinetic friction* (or *sliding friction*) that occurs afterward is usually less than the starting friction, so less force is needed to keep the book moving than to start it moving (Figure 3-4).

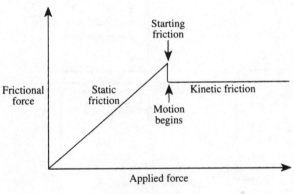

Figure 3-4

Coefficient of Friction

The frictional force between two surfaces depends on the normal (perpendicular) force N pressing them together and on the nature of the surfaces. The latter factor is expressed quantitatively in the *coefficient of friction* μ (Greek letter *mu*) whose value depends on the materials in contact. The frictional force is experimentally found to be:

$$F_f \leq \mu_s N \quad \text{Static friction}$$

$$F_f = \mu_k N \quad \text{Kinetic friction}$$

In the case of static friction, F_f increases as the applied force increases until the limiting value of $\mu_s N$ is reached. Thus when there is no motion, $\mu_s N$ gives the starting frictional force, not the actual frictional force. Up to $\mu_s N$, the actual frictional force F_f has the same magnitude as the applied force but is in the opposite direction.

When the applied force exceeds the starting frictional force $\mu_s N$, motion begins and now the coefficient of kinetic friction μ_k governs the frictional force. In this case, $\mu_k N$ gives the actual amount of F_f, which no longer depends on the applied force and is constant over a fairly wide range of relative velocities.

Solved Problem 3.4 A force of 200 N is just sufficient to start a 50-kg steel trunk moving across a wooden floor. Find the coefficient of static friction.

Solution. The normal force is the trunk's weight mg. Hence,

$$\mu_s = \frac{F}{N} = \frac{F}{mg} = \frac{200 \text{ N}}{(50 \text{ kg})(9.8 \text{ m}/\text{s}^2)} = 0.41$$

Chapter 4
ENERGY

Work

Work is a measure of the amount of change (in a general sense) that a force produces when it acts on a body. The change may be in the velocity of the body, in its position, or in its size or shape.

By definition, the work done by a force acting on a body is equal to the product of the force and the distance through which the force acts, provided that **F** and **s** are in the same direction. Thus

$$W = Fs$$
$$\text{Work} = (\text{force})(\text{distance})$$

Work is a scalar quantity; no direction is associated with it.

If **F** and **s** are not parallel but **F** is at the angle θ with respect to **s**, then

28

$$W = Fs \cos \theta$$

Since $\cos 0° = 1$, this formula becomes $W = Fs$ when **F** is parallel to **s**. When **F** is perpendicular to **s**, $\theta = 90°$ and $\cos 90° = 0$. No work is done in this case (Figure 4-1).

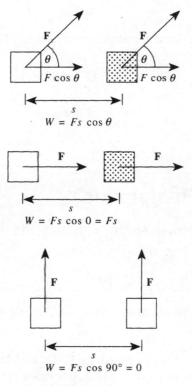

Figure 4-1

The unit of work is the product of a force unit and a length unit. In SI units, the unit of work is the *joule* (J).

SI units: 1 joule (J) = 1 newton-meter = 0.738 ft·lb

Solved Problem 4.1 A horizontal force of 420 N is used to push a 100-kg crate for 5 m across a level warehouse floor. How much work is done?

Solution. The mass of the crate does not matter here. Since the force is parallel to the displacement,

$$W = Fs = (420 \text{ N})(5 \text{ m}) = 2100 \text{ J} = 2.1 \text{ kJ}$$

Power

Power is the rate at which work is done by a force. Thus

$$P = \frac{W}{t}$$

$$\text{Power} = \frac{\text{work done}}{\text{time}}$$

Remember

The more power something has, the more work it can perform in a given time.

Two special units of power are in wide use, the *watt* and the *horsepower*, where

$$1 \text{ watt (W)} = 1 \text{ J/s} = 1.34 \times 10^{-3} \text{ hp}$$
$$1 \text{ horsepower (hp)} = 550 \text{ ft·lb/s} = 746 \text{ W}$$

When a constant force **F** does work on a body that is moving at the constant velocity **v**, if **F** is parallel to **v**, the power involved is

$$P = \frac{W}{t} = \frac{Fs}{t} = Fv$$

because $s/t = v$; that is

$$P = Fv$$

$$\text{Power} = (\text{force})(\text{velocity})$$

Solved Problem 4.2 A 40-kg woman runs up a staircase 4 m high in 5 s. Find her minimum power output.

Solution. The minimum downward force the woman's legs must exert is equal to her weight mg. Hence

$$P = \frac{W}{t} = \frac{Fs}{t} = \frac{mgh}{t} = \frac{(40 \text{ kg})(9.8 \text{ m/s}^2)(4 \text{ m})}{5 \text{ s}} = 314 \text{ W}$$

Kinetic Energy

Energy is that property something has that enables it to do work. The more energy something has, the more work it can perform. Two general categories of energy are kinetic energy and potential energy.

☆ Note!

The units of energy are the same as those of work, namely the joule and the foot-pound.

The energy a body has by virtue of its motion is called *kinetic energy*. If the body's mass is m and its velocity is v, its kinetic energy is

$$\text{Kinetic energy} = \text{KE} = \frac{1}{2}mv^2$$

Solved Problem 4.3 Find the kinetic energy of a 1000-kg car whose velocity is 20 m/s.

Solution. $\text{KE} = \frac{1}{2}mv^2 = \frac{1}{2}(1000 \text{ kg})(20 \text{ m/s})^2 = 2 \times 10^5 \text{ J}$

Potential Energy

The energy a body has by virtue of its position is called *potential energy*. A book held above the floor has gravitational potential energy because the book can do work on something else as it falls; a nail held near a magnet has magnetic potential energy because the nail can do work as it moves toward the magnet; the wound spring in a watch has elastic potential energy because the spring can do work as it unwinds.

The gravitational potential energy of a body of mass m at a height h above a given reference level is:

$$\text{Gravitational potential energy} = PE = mgh$$

where g is the acceleration due to gravity.

Solved Problem 4.4 A 1.5-kg book is held 60 cm above a desk whose top is 70 cm above the floor. Find the potential energy of the book (*a*) with respect to the desk, and (*b*) with respect to the floor.

Solution.

(*a*) Here $h = 60$ cm $= 0.6$ m, so

$$PE = mgh = (1.5 \text{ kg})(9.8 \text{ m/s}^2)(0.6 \text{ m}) = 8.8 \text{ J}$$

(*b*) The book is $h = 60$ cm $+ 70$ cm $= 130$ cm $= 1.3$ m above the floor, so its PE with respect to the floor is

$$PE = mgh = (1.5 \text{ kg})(9.8 \text{ m/s}^2)(1.3 \text{ m}) = 19.1 \text{ J}$$

Conservation of Energy

According to the law of *conservation of energy*, energy cannot be created or destroyed, although it can be transformed from one kind to another. The total amount of energy in the universe is constant. A falling stone provides a simple example: More and more of its initial potential energy turns to kinetic energy as its velocity increases, until finally all its PE has become KE when it strikes the ground. The KE of the stone is then transferred to the ground as work by the impact.

In general,

Work done *on* an object = change in object's KE + change
in object's PE + work done *by* object

Work done by an object against friction becomes heat.

Chapter 5
MOMENTUM

IN THIS CHAPTER:

✔ *Linear Momentum*
✔ *Impulse*
✔ *Conservation of Linear Momentum*
✔ *Collisions*

Linear Momentum

Work and energy are scalar quantities that have no directions associated with them. When two or more bodies interact with one another, or a single body breaks
up into two or more new bodies, the various directions
of motion cannot be related by energy considerations
alone. The vector quantities, called *linear momentum*
and *impulse*, are important in analyzing such events.

The linear momentum (usually called simply *momentum*) of a body
of mass m and velocity $\mathbf{v}$ is the product of m and $\mathbf{v}$:

$$\text{Momentum} = m\mathbf{v}$$

The units of momentum are kilogram-meters per second and slug-feet per
second. The direction of the momentum of a body is the same as the direction in which it is moving.

The greater the momentum of a body, the greater its tendency to continue in motion. Thus, a baseball that is solidly struck by a bat (v large) is harder to stop than a baseball thrown by hand (v small), and an iron shot (m large) is harder to stop than a baseball (m small) of the same velocity.

Solved Problem 5.1 Find the momentum of a 50-kg boy running at 6 m/s.

Solution. The momentum can be calculated as follows:

$$mv = (50 \text{ kg})(6 \text{ m/s}) = 300 \text{ kg·m/s}$$

Impulse

A force $\mathbf{F}$ that acts on a body during time t provides the body with an *impulse* of $\mathbf{F}t$:

$$\text{Impulse} = \mathbf{F}t = (\text{force})(\text{time interval})$$

You Need to Know

The units of impulse are newton-seconds and pound-seconds.

When a force acts on a body to produce a change in its momentum, the momentum change $m(\mathbf{v}_2 - \mathbf{v}_1)$ is equal to the impulse provided by the force. Thus

$$\mathbf{F}t = m(\mathbf{v}_2 - \mathbf{v}_1)$$

Impulse = momentum change

Solved Problem 5.2 A 46-g golf ball is struck by a club and flies off at 70 m/s. If the head of the club was in contact with the ball for 0.5 ms, what was the average force on the ball during the impact?

Solution. The ball started from rest, so $v_1 = 0$ and its momentum change is:

$$m(v_2 - v_1) = mv_2 = (0.046 \text{ kg})(70 \text{ m/s}) = 3.22 \text{ kg·m/s}$$

Since 1 ms = 1 millisecond = 10^{-3} s, here $t = 0.5$ ms $= 5 \times 10^{-4}$ s and

$$F = \frac{m(v_2 - v_1)}{t} = \frac{3.22 \text{ kg·m/s}}{5 \times 10^{-4} \text{ s}} = 6.4 \times 10^3 \text{ N} = 6.4 \text{ kN}$$

Conservation of Linear Momentum

According to the law of *conservation of linear momentum*, when the vector sum of the external forces that act on a system of bodies equals zero, the total linear momentum of the system remains constant no matter what momentum changes occur within the system.

Although interactions within the system may change the *distribution* of the total momentum among the various bodies in the system, the total momentum does not change. Such interactions can give rise to two general classes of events: explosions, in which an original single body flies apart into separate bodies, and collisions, in which two or more bodies collide and either stick together or move apart, in each case with a redistribution of the original total momentum.

Solved Problem 5.3 A rocket explodes in midair. How does this affect (*a*) its total momentum and (*b*) its total kinetic energy?

Solution.

(*a*) The total momentum remains the same because no external forces acted on the rocket.

(*b*) The total kinetic energy increases because the rocket fragments received additional KE from the explosion.

Collisions

Momentum is also conserved in collisions. If a moving billiard ball strikes a stationary one, the two move off in such a way that the vector sum of their momenta is the same as the initial momentum of the first ball (Figure 5-1). This is true even if the balls move off in different directions.

A perfectly *elastic* collision is one in which the bodies involved move apart in such a way that kinetic energy as well as momentum is conserved. In a perfectly *inelastic* collision, the bodies stick together and the kinetic energy loss is the maximum possible consistent with momentum conservation. Most collisions are intermediate between these two extremes.

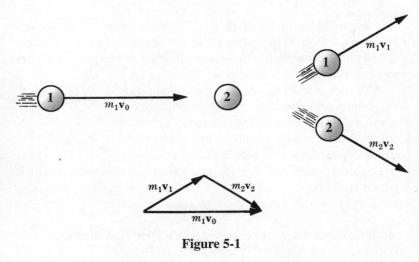

Figure 5-1

Solved Problem 5.4 A 2000-lb car moving at 50 mi/h collides head-on with a 3000-lb car moving at 20 mi/h, and the two cars stick together. Which way does the wreckage move?

Solution. The 2000-lb car had the greater initial momentum, so the wreckage moves in the same direction it had.

Chapter 6
CIRCULAR MOTION AND GRAVITATION

IN THIS CHAPTER:

- ✔ *Centripetal Acceleration*
- ✔ *Centripetal Force*
- ✔ *Motion in a Vertical Circle*
- ✔ *Gravitation*
- ✔ *Satellite Motion*

Centripetal Acceleration

A body that moves in a circular path with a velocity whose magnitude is constant is said to undergo *uniform circular motion.*

Although the velocity of a body in uniform circular motion is constant in magnitude, its direction changes continually. The body is therefore accelerated. The direction of this *centripetal acceleration* is toward the center of the circle in which the body moves, and its magnitude is

$$a_c = \frac{v^2}{r}$$

$$\text{Centripetal acceleration} = \frac{(\text{velocity of body})^2}{\text{radius of circular path}}$$

Note!

Because the acceleration is perpendicular to the path followed by the body, the body's velocity changes only in direction, not in magnitude.

Centripetal Force

The inward force that must be applied to keep a body moving in a circle is called *centripetal force*. Without centripetal force, circular motion cannot occur. Since $F = ma$, the magnitude of the centripetal force on a body in uniform circular motion is

$$\text{Centripetal force } = F_c = \frac{mv^2}{r}$$

Solved Problem 6.1 A 1000-kg car rounds a turn of radius 30 m at a velocity of 9 m/s. (*a*) How much centripetal force is required? (*b*) Where does this force come from?

Solution.

(*a*) $\quad F_c = \dfrac{mv^2}{r} = \dfrac{(1000 \text{ kg})(9 \text{ m/s})^2}{30 \text{ m}} = 2700 \text{ N}$

(*b*) The centripetal force on a car making a turn on a level road is provided by the road acting via friction on the car's tires.

Motion in a Vertical Circle

When a body moves in a vertical circle at the end of a string, the tension **T** in the string varies with the body's position. The centripetal force **F**$_c$ on the body at any point is the vector sum of **T** and the component of the body's weight **w** toward the center of the circle. At the top of the circle, as in Figure 6-1(*a*), the weight **w** and the tension **T** both act toward the center of the circle, and so

$$T = F_c - w$$

At the bottom of the circle, as in Figure 6-1(b), **w** acts away from the center of the circle, and so

$$T = F_c + w$$

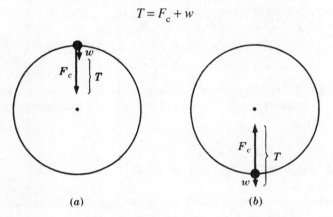

(a)   (b)

Figure 6-1

Solved Problem 6.2 A string 0.5 m long is used to whirl a 1-kg stone in a vertical circle at a uniform velocity of 5 m/s. What is the tension of the string (a) when the stone is at the top of the circle and (b) when the stone is at the bottom of the circle?

Solution.
(a) The centripetal force needed to keep the stone moving at 5 m/s is

$$F_c = \frac{mv^2}{r} = \frac{(1 \text{ kg})(5 \text{ m/s})^2}{0.5 \text{ m}} = 50 \text{ N}$$

The weight of the stone is $w = mg = (1 \text{ kg})(9.8 \text{ m/s}^2) = 9.8$ N. At the top of the circle,

$$T = F_c - w = 50 \text{ N} - 9.8 \text{ N} = 40.2 \text{ N}$$

(b) At the bottom of the circle,

$$T = F_c + w = 59.8 \text{ N}$$

Gravitation

According to Newton's *law of universal gravitation*, every body in the universe attracts every other body with a force that is directly proportional to each of their masses and inversely proportional to the square of the distance between them. In equation form,

$$\text{Gravitational force} = F_g = G\frac{m_1 m_2}{r^2}$$

where m_1 and m_2 are the masses of any two bodies, r is the distance between them, and G is a constant whose values in SI and British units are, respectively,

SI units: $G = 6.67 \times 10^{-11}$ N $\cdot$ m^2/kg^2
British units: $G = 3.34 \times 10^{-8}$ lb $\cdot$ ft^2/slug^2

A spherical body behaves gravitationally as though its entire mass were concentrated at its center.

Solved Problem 6.3 What gravitational force does a 1000-kg lead sphere exert on an identical sphere 3 m away?

Solution.

$$F_g = G\frac{m_1 m_2}{r^2} = \frac{\left(6.67 \times 10^{-11} \text{ N} \cdot \text{m}^2 / \text{kg}^2\right)\left(10^3 \text{ kg}\right)\left(10^3 \text{ kg}\right)}{(3 \text{ m})^2}$$

$$= 7.4 \times 10^{-4} \text{ N}$$

This is less than the force that would result from blowing gently on one of the spheres. Gravitational forces are usually significant only when at least one of the bodies has a very large mass.

Satellite Motion

Gravitation provides the centripetal forces that keep the planets in their orbits around the sun and the moon in its orbit around the earth. The same is true for artificial satellites put into orbit around the earth.

Chapter 7
ROTATIONAL MOTION

IN THIS CHAPTER:

- ✔ *Angular Measure*
- ✔ *Angular Velocity*
- ✔ *Angular Acceleration*
- ✔ *Moment of Inertia*
- ✔ *Torque*
- ✔ *Rotational Energy and Work*
- ✔ *Angular Momentum*

Angular Measure

In everyday life, angles are measured in degrees, where 360° equals a full turn. A more suitable unit for technical purposes is the *radian* (rad). If a circle is drawn whose center is at the vertex of a particular angle (Figure 7-1), the angle θ (Greek letter *theta*) in radians is equal to the ratio between the arc *s* cut by the angle and the radius *r* of the circle:

$$\theta = \frac{s}{r}$$

$$\text{Angle in radians} = \frac{\text{arc length}}{\text{radius}}$$

41

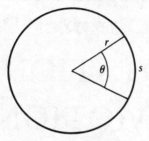

Figure 7-1

Because the circumference of a circle of radius r is $2\pi r$, there are 2π rad in a complete revolution (rev). Hence

$$1 \text{ rev} = 360° = 2\pi \text{ rad}$$

and so,

$$1° = 0.01745 \text{ rad} \qquad 1 \text{ rad} = 57.30°$$

Angular Velocity

The *angular velocity* of a body describes how fast it is turning about an axis. If a body turns through the angle θ in the time t, its angular velocity ω (Greek letter *omega*) is

$$\omega = \frac{\theta}{t}$$

$$\text{Angular velocity} = \frac{\text{angular displacement}}{\text{time}}$$

Angular velocity is usually expressed in radians per second (rad/s), revolutions per second (rev/s or rps), and revolutions per minute (rev/min or rpm), where

$$1 \text{ rev/s} = 2\pi \text{ rad/s} = 6.28 \text{ rad/s}$$

$$1 \text{ rev/min} = \frac{2\pi}{60} \text{ rad/s} = 0.105 \text{ rad/s}$$

The linear velocity v of a particle that moves in a circle of radius r with the uniform angular velocity ω is given by

$$v = \omega r$$

Linear velocity $=$ (angular velocity)(radius of circle)

This formula is valid only when ω is expressed in radian measure.

Angular Acceleration

A rotating body whose angular velocity changes from ω_0 to ω_f in the time interval t has the *angular acceleration* α (Greek letter *alpha*) of

$$\alpha = \frac{\omega_f - \omega_o}{t}$$

Angular acceleration $= \dfrac{\text{angular velocity change}}{\text{time}}$

A positive value of α means that the angular velocity is increasing; a negative value means that it is decreasing. Only constant angular accelerations are considered here.

The formulas relating the angular displacement, velocity, and acceleration of a rotating body under constant angular acceleration are analogous to the formulas relating linear displacement, velocity, and acceleration. If a body has the initial angular velocity ω_0, its angular velocity ω_f after a time t during which its angular acceleration is α will be

$$\omega_f = \omega_o + \alpha t$$

and, in this time, it will have turned through an angular displacement of

$$\theta = \omega_o t + \frac{1}{2}\alpha t^2$$

A relationship that does not involve the time t directly is sometimes useful:

$$\omega_f^2 = \omega_o^2 + 2\alpha\theta$$

Solved Problem 7.1 A phonograph turntable initially rotating at 3.5 rad/s makes three complete turns before coming to a stop. (*a*) What is its angular acceleration? (*b*) How much time does it take to come to a stop?

Solution.

(*a*) The angle in radians that corresponds to 3 rev is

$$\theta = (3 \text{ rev})(2\pi \text{ rad/rev}) = 6\pi \text{ rad}$$

From the formula, $\omega_f^2 = \omega_o^2 + 2\alpha\theta$, we find

$$\alpha = \frac{\omega_f^2 - \omega_o^2}{2\theta} = \frac{0 - (3.5 \text{ rad/s})^2}{(2)(6\pi \text{ rad})} = -0.325 \text{ rad/s}^2$$

(*b*) Since $\omega_f = \omega_o + \alpha t$, we have here

$$t = \frac{\omega_f - \omega_o}{\alpha} = \frac{0 - 3.5 \text{ rad/s}}{-0.325 \text{ rad/s}^2} = 10.8 \text{ s}$$

Moment of Inertia

The rotational analog of mass is a quantity called *moment of inertia*. The greater the moment of inertia of a body, the greater its resistance to a change in its angular velocity.

You Need to Know

The value of the moment of inertia I of a body about a particular axis of rotation depends not only upon the body's mass but also upon how the mass is distributed about the axis.

Let us imagine a rigid body divided into a great many small particles whose masses are $m_1, m_2, m_3, \ldots$ and whose distances from the axis of rotation are respectively $r_1, r_2, r_3, \ldots$ (Figure 7-2).

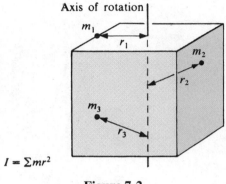

Axis of rotation

$I = \Sigma mr^2$

Figure 7-2

The moment of inertia of this body is given by

$$I = m_1 r_1^2 + m_2 r_2^2 + m_3 r_3^2 + \cdots = \Sigma mr^2$$

where the symbol Σ (Greek capital letter *sigma*) means "sum of" as before. The farther a particle is from the axis of rotation, the more it contributes to the moment of inertia. The units of I are kg $\cdot$ m^2 and slug $\cdot$ ft^2. Some examples of moments of inertia of bodies of mass M are shown in Figure 7-3.

Torque

The *torque* τ (Greek letter *tau*) exerted by a force on a body is a measure of its effectiveness in turning the body about a certain pivot point. The *moment arm* of a force **F** about a pivot point O is the perpendicular distance L between the line of action of the force and O (Figure 7-4). The torque τ exerted by the force about O has the magnitude

$$\tau = FL$$

Torque $=$ (force)(moment arm)

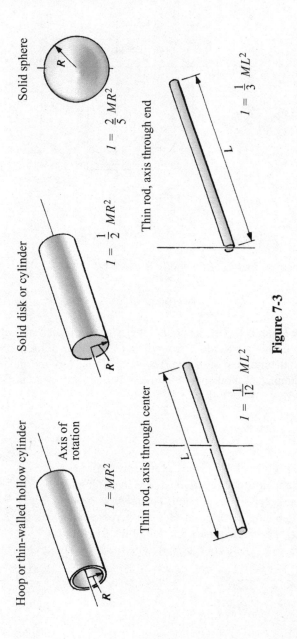

Figure 7-3

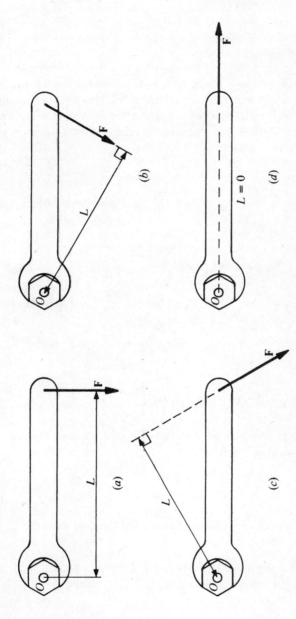

Figure 7-4 Four directions along which a force **F** can be applied to a wrench. In (*a*) the moment arm *L* is longest, hence the torque $\tau = FL$ is a maximum. In (*d*) the line of action of **F** passes through the pivot *O*, so $L = 0$ and $\tau = 0$.

The torque exerted by a force is also known as the *moment* of the force. A force whose line of action passes through O produces no torque about O because its moment arm is zero.

Torque plays the same role in rotational motion that force plays in linear motion. A net force F acting on a body of mass m causes it to undergo the linear acceleration a in accordance with Newton's second law of motion $F = ma$. Similarly, a net torque τ acting on a body of moment of inertia I causes it to undergo the angular acceleration α (in rad/s^2) in accordance with the formula

$$\tau = I\alpha$$

Torque = (moment of inertia) (angular acceleration)

Remember

In the SI system, the unit of torque is newton · meter (N·m); in the British system, it is the pound · foot (lb·ft).

Rotational Energy and Work

The kinetic energy of a body of moment of inertia I whose angular velocity is ω (in rad/s) is

$$KE = \frac{1}{2} I\omega^2$$

Kinetic energy = $\left(\frac{1}{2}\right)$(moment of inertia) (angular velocity)^2

The work done by a constant torque τ that acts on a body while it experiences the angular displacement θ rad is

$$W = \tau\theta$$

Work = (torque) (angular displacement)

The rate at which work is being done when a torque τ acts on a body that rotates at the constant angular velocity ω (rad/s) is

$$P = \tau\omega$$

Power $=$ (torque) (angular velocity)

Angular Momentum

The equivalent of linear momentum in rotational motion is *angular momentum*. The angular momentum **L** of a rotating body has the magnitude

$$L = I\omega$$

Angular momentum $=$ (moment of inertia) (angular velocity)

The greater the angular momentum of a spinning object, such as a top, the greater its tendency to spin.

Like linear momentum, angular momentum is a vector quantity with direction as well as magnitude. The direction of the angular momentum of a rotating body is given by the right-hand rule (Figure 7-5):

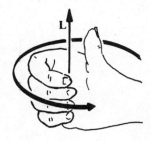

Figure 7-5

When the fingers of the right hand are curled in the direction of rotation, the thumb points in the direction of **L**.

According to the principle of *conservation of angular momentum*, the total angular momentum of a system of bodies remains constant in the absence of a net torque regardless of what happens within the system. Because angular momentum is a vector quantity, its conservation implies that the direction of the axis of rotation tends to remain unchanged.

Table 7.1 compares linear and angular quantities.

Linear Quantity		Angular Quantity	
Distance	$s = v_0 t + \frac{1}{2} at^2$ $s = \left(\dfrac{v_0 + v_f}{2} \right) t$	Angle	$\theta = \omega_0 t + \frac{1}{2} \alpha t^2$ $\theta = \left(\dfrac{\omega_0 + \omega_f}{2} \right) t$
Speed	$v = v_0 + at$ $v^2 = v_0^2 + 2as$	Angular speed	$\omega = \omega_0 + \alpha t$ $\omega^2 = \omega_0^2 + 2\alpha\theta$
Acceleration	$a = \Delta v / \Delta t$	Angular acceleration	$\alpha = \Delta \omega / \Delta t$
Mass	m	Moment of inertia	I
Force	$F = ma$	Torque	$\tau = I\alpha$
Momentum	$p = mv$	Angular momentum	$L = I\omega$
Work	$W = Fs$	Work	$W = \tau\theta$
Power	$P = Fv$	Power	$P = \tau\omega$
Kinetic energy	$KE = \frac{1}{2} mv^2$	Kinetic energy	$KE = \frac{1}{2} I\omega^2$

Table 7.1

Chapter 8
EQUILIBRIUM

IN THIS CHAPTER:

✔ *Translational Equilibrium*
✔ *Rotational Equilibrium*
✔ *Center of Gravity*
✔ *Finding a Center of Gravity*

Translational Equilibrium

A body is in *translational equilibrium* when no net force acts on it. Such a body is not accelerated, and it remains either at rest or in motion at constant velocity along a straight line, whichever its initial state was.

A body in translational equilibrium may have forces acting on it, but they must be such that their vector sum is zero. Thus the condition for the translational equilibrium of a body may be written

$$\Sigma \mathbf{F} = 0$$

where the symbol Σ (Greek capital letter *sigma*) means "sum of" and $\mathbf{F}$ refers to the various forces that act on the body.

The procedure for working out a problem that involves translational equilibrium has three steps:

1. Draw a diagram of the forces that act *on* the body. This is called a *free-body diagram*.
2. Choose a set of coordinate axes and resolve the various forces into their components along these axes.
3. Set the sum of the force components along each axis equal to zero so that

$$\text{Sum of } x \text{ force components} = \Sigma F_x = 0$$
$$\text{Sum of } y \text{ force components} = \Sigma F_y = 0$$
$$\text{Sum of } z \text{ force components} = \Sigma F_z = 0$$

In this way, the vector equation $\Sigma \mathbf{F} = 0$ is replaced by three scalar equations. Then solve the resulting equations for the unknown quantities.

A proper choice of directions for the axes often simplifies the calculations. When all the forces lie in a plane, for instance, the coordinate system can be chosen so that the x and y axes lie in the plane; then the two equations $\Sigma F_x = 0$ and $\Sigma F_y = 0$ are enough to express the condition for translational equilibrium.

Solved Problem 8.1 A 100-N box is suspended from two ropes that each make an angle of $40°$ with the vertical. Find the tension in each rope.

Solution. The forces that act on the box are shown in the free-body diagram of Figure 8-1(*a*). They are

$$\mathbf{T}_1 = \text{tension in left-hand rope}$$
$$\mathbf{T}_2 = \text{tension in right-hand rope}$$
$$\mathbf{w} = \text{weight of box, which acts downward}$$

Since the forces all lie in a plane, we need only x and y axes. In Figure 8-1(*b*), the forces are resolved into their x and y components, whose magnitudes are as follows:

$$T_{1x} = -T_1 \sin \theta_1 = -T_1 \sin 40° = -0.643 T_1$$
$$T_{1y} = T_1 \cos \theta_1 = T_1 \cos 40° = 0.766 T_1$$
$$T_{2x} = T_2 \sin \theta_2 = T_2 \sin 40° = 0.643 T_2$$
$$T_{2y} = T_2 \cos \theta_2 = T_2 \cos 40° = 0.766 T_2$$
$$w = -100 \text{ N}$$

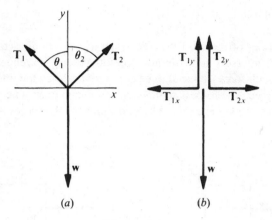

Figure 8-1

Because $\mathbf{T}_{1x}$ and $\mathbf{w}$ are, respectively, in the $-x$ and $-y$ directions, both have negative magnitudes.

Now we are ready for step 3. First we add the x components of the forces and set the sum equal to zero. This yields

$$\Sigma F_x = T_{1x} + T_{2x} = 0$$
$$-0.643T_1 + 0.643T_2 = 0$$
$$T_1 = T_2 = T$$

Evidently, the tensions in the two ropes are equal. Next we do the same for the y components:

$$\Sigma F_y = T_{1y} + T_{2y} + w = 0$$
$$0.766T_1 + 0.766T_2 - 100 \text{ N} = 0$$
$$0.766(T_1 + T_2) = 100 \text{ N}$$
$$T_1 + T_2 = \frac{100 \text{ N}}{0.766} = 130.5 \text{ N}$$

Since $T_1 = T_2 = T$,

$$T_1 = T_2 = 2T = 130.5 \text{ N}$$
$$T = 65 \text{ N}$$

The tension in each rope is 65 N.

Rotational Equilibrium

When the lines of action of the forces that act on a body in translational equilibrium intersect at a common point, they have no tendency to turn the body. Such forces are said to be *concurrent*. When the lines of action do not intersect, the forces are *nonconcurrent* and exert a net torque that acts to turn the body even though the resultant of the forces is zero (Figure 8-2).

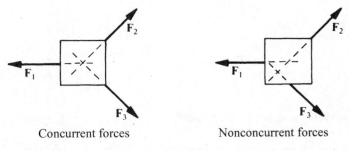

Concurrent forces　　　　　Nonconcurrent forces

Figure 8-2

A body is in *rotational equilibrium* when no net torque acts on it. Such a body remains in its initial rotational state, either not spinning at all or spinning at a constant rate. The condition for the rotational equilibrium of a body may therefore be written

$$\Sigma \tau = 0$$

where $\Sigma \tau$ refers to the sum of the torques acting on the body about any point.

A torque that tends to cause a counterclockwise rotation when it is viewed from a given direction is considered positive; a torque that tends to cause a clockwise rotation is considered negative (Figure 8-3).

To investigate the rotational equilibrium of a body, any convenient point may be used as the pivot point for calculating torques; if the sum of the torques on a body in translational equilibrium is zero about some point, it is zero about any other point.

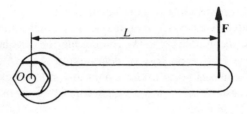

$$\tau = +FL$$

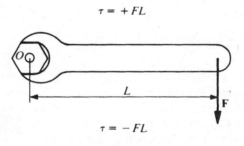

$$\tau = -FL$$

Figure 8-3

Center of Gravity

The *center of gravity* of a body is that point at which the body's entire weight can be regarded as being concentrated. A body can be suspended in any orientation from its center of gravity without tending to rotate.

★ Note!

In analyzing the equilibrium of a body, its weight can be considered as a downward force acting from its center of gravity.

Solved Problem 8.2 (*a*) Under what circumstances is it necessary to consider torques in analyzing an equilibrium situation? (*b*) About what point should torques be calculated when this is necessary?

Solution. (*a*) Torques must be considered when the various forces that act on the body are nonconcurrent, that is, when their lines of action do not intersect at a common point. (*b*) Torques may be calculated about any point whatever for the purpose of determining the equilibrium of the body. Hence it makes sense to use a point that minimizes the labor involved, which usually is the point through which pass the maximum number of lines of action of the various forces; this is because a force whose line of action passes through a point exerts no torque about that point.

Finding a Center of Gravity

The center of gravity (CG) of an object of regular form and uniform composition is located at its geometric center. In the case of a complex object, the way to find its center of gravity is to consider it as a system of separate particles and then find the balance point of the system. An example is the massless rod of Figure 8-4, which has three particles m_1, m_2, and m_3 attached to it.

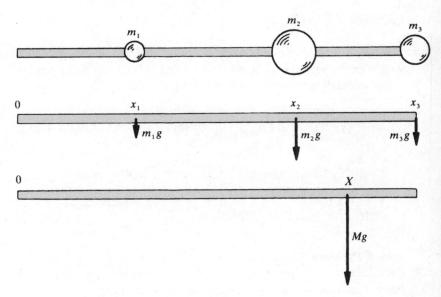

Figure 8-4

The CG of the system is at a distance X from the end of the rod such that the torque exerted by a single particle of mass $M = m_1 + m_2 + m_3$ at X equals the sum of the torques exerted by the particles at their locations x_1, x_2, and x_3. Thus,

$$m_1 g x_1 + m_2 g x_2 + m_3 g x_3 = MgX = (m_1 + m_2 + m_3)gX$$

$$X = \frac{m_1 x_1 + m_2 x_2 + m_3 x_3}{m_1 + m_2 + m_3}$$

This formula can be extended to any number of particles. If the complex object involves two or three dimensions rather than just one, the same procedure is applied along two or three coordinate axes to find X and Y or X, Y, and Z, which are the coordinates of the center of gravity.

Chapter 9
SIMPLE HARMONIC MOTION

IN THIS CHAPTER:

✔ *Restoring Force*
✔ *Elastic Potential Energy*
✔ *Simple Harmonic Motion*
✔ *Period and Frequency*
✔ *Displacement, Velocity, and Acceleration*
✔ *Pendulums*

Restoring Force

When an elastic object such as a spring is stretched or compressed, a *restoring force* appears that tries to return the object to its normal length. It is this restoring force that must be overcome by the applied force in order to deform the object. From Hooke's law, the restoring force F_r is proportional to the displacement s provided the elastic limit is not exceeded. Hence

$$F_r = -ks$$

Restoring force $= -$(force constant) (displacement)

The minus sign is required because the restoring force acts in the opposite direction to the displacement. The greater the value of the *force constant* k, the greater the restoring force for a given displacement and the greater the applied force $F = ks$ needed to produce the displacement.

Elastic Potential Energy

Because work must be done by an applied force to stretch or compress an object, the object has *elastic potential energy*, where

$$PE = \frac{1}{2}ks^2$$

When a deformed elastic object is released, its elastic potential energy turns into kinetic energy or into work done on something else.

Solved Problem 9.1 A force of 5 N compresses a spring by 4 cm. (*a*) Find the force constant of the spring. (*b*) Find the elastic potential energy of the compressed spring.

Solution.

(*a*) $$k = \frac{F}{s} = \frac{5\ \text{N}}{0.04\ \text{m}} = 125\ \text{N/m}$$

(*b*) $$PE = \frac{1}{2}ks^2 = \left(\frac{1}{2}\right)(125\ \text{N/m})(0.04\ \text{m})^2 = 0.1\ \text{J}$$

Simple Harmonic Motion

In *periodic motion*, a body repeats a certain motion indefinitely, always returning to its starting point after a constant time interval and then starting a new cycle. *Simple harmonic motion* is periodic motion that occurs when the restoring force on a body displaced from an equilibrium position is proportional to the displacement and in the opposite direction. A mass m attached to a spring executes simple harmonic motion when the spring is pulled out and released. The spring's PE becomes KE as the

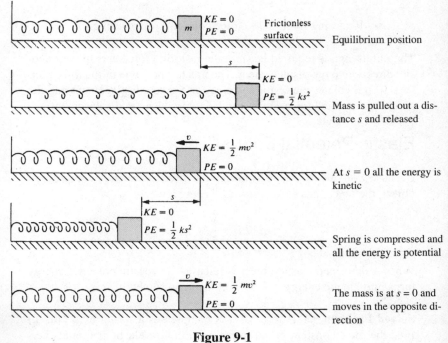

Figure 9-1

mass begins to move, and the KE of the mass becomes PE again as its momentum causes the spring to overshoot the equilibrium position and become compressed (Figure 9-1).

The *amplitude A* of a body undergoing simple harmonic motion is the maximum value of its displacement on either side of the equilibrium position.

Period and Frequency

The *period T* of a body undergoing simple harmonic motion is the time needed for one complete cycle; *T* is independent of the amplitude *A*. If the acceleration of the body is *a* when its displacement is *s*,

$$T = 2\pi \sqrt{-\frac{s}{a}}$$

$$\text{Period} = 2\pi \sqrt{-\frac{\text{displacement}}{\text{acceleration}}}$$

In the case of a body of mass m attached to a spring of force constant k, $F_r = -ks = ma$, and so $-s/a = m/k$. Hence

$$T = 2\pi \sqrt{\frac{m}{k}} \quad \text{stretched spring}$$

The *frequency f* of a body undergoing simple harmonic motion is the number of cycles per second it executes, so that

$$f = \frac{1}{T}$$

$$\text{Frequency} = \frac{1}{\text{period}}$$

The unit of frequency is the *hertz* (Hz), where 1 Hz = 1 cycle/s.

Displacement, Velocity, and Acceleration

If $t = 0$ when a body undergoing simple harmonic motion is in its equilibrium position of $s = 0$ and is moving in the direction of increasing s, then at any time t thereafter its displacement is

$$s = A \sin 2\pi f t$$

Often, this formula is written

$$s = A \sin \omega t$$

where $\omega = 2\pi f$ is the *angular frequency* of the motion in radians per second. Figure 9-2 is a graph of s versus t.

The velocity of the body at time t is

$$v = 2\pi f A \cos 2\pi f t = \omega A \cos \omega t$$

When v is positive, the body is moving in the direction of increasing s; when v is negative, it is moving in the direction of decreasing s. In terms of the displacement s, the magnitude of the velocity is

$$v = 2\pi f \sqrt{A^2 - s^2}$$

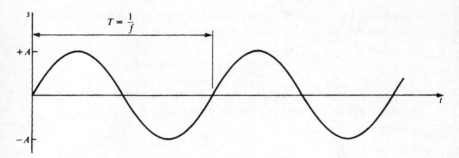

Figure 9-2

The acceleration of the body at time t is

$$a = -4\pi^2 f^2 A \sin 2\pi f t = -\omega^2 A \sin \omega t$$

In terms of the displacement s, the acceleration is

$$a = 4\pi^2 f^2 s$$

Pendulums

A *simple pendulum* has its entire mass concentrated at the end of the string, as in Figure 9-3(a), and it undergoes simple harmonic motion pro-

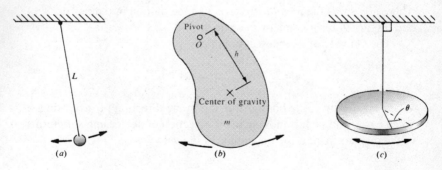

Figure 9-3

vided that the arc through which it travels is only a few degrees. The period of a simple pendulum of length L is

$$T = 2\pi \sqrt{\frac{L}{g}} \quad \text{simple pendulum}$$

The *physical pendulum* of Figure 9-3(*b*) is an object of any kind which is pivoted so that it can oscillate freely. If the moment of inertia of the object about the pivot O is I, its mass is m, and the distance from its center of gravity to the pivot is h, then its period is

$$T = 2\pi \sqrt{\frac{I}{mgh}} \quad \text{physical pendulum}$$

A *torsion pendulum* consists of an object suspended by a wire or thin rod, as in Figure 9-3(*c*), which undergoes rotational simple harmonic oscillations. From Hooke's law, the torque τ needed to twist the object through an angle θ is

$$\tau = K\theta$$

provided the elastic limit is not exceeded, where K is a constant that depends on the material and dimensions of the wire. If I is the moment of inertia of the object about its point of suspension, the period of the oscillation is

$$T = 2\pi \sqrt{\frac{I}{K}} \quad \text{torsion pendulum}$$

Solved Problem 9.2 A lamp is suspended from a high ceiling with a cord 12 ft long. Find its period of oscillation.

Solution.

$$T = 2\pi \sqrt{\frac{L}{g}} = 2\pi \sqrt{\frac{12 \text{ ft}}{32 \text{ ft} / \text{s}^2}} = 3.85 \text{ s}$$

IN THIS CHAPTER:

- ✔ *Waves*
- ✔ *Wave Properties*
- ✔ *Logarithms*
- ✔ *Sound*
- ✔ *Doppler Effect*

Waves

A *wave* is, in general, a disturbance that moves through a medium. An exception is an *electromagnetic wave*, which can travel through a vacuum. Examples are light and radio waves. A wave carries energy, but there is no transport of matter. In a *periodic wave*, pulses of the same kind follow one another in regular succession.

In a *transverse wave*, the particles of the medium move back and forth perpendicular to the direction of the wave. Waves that travel down a stretched string when one end is shaken are transverse (Figure 10-1).

In a *longitudinal wave*, the particles of the medium move back and forth in the same direction as the wave. Waves that travel down a coil spring when one end is pulled out and released are longitudinal (Figure 10-2). Sound waves are also longitudinal.

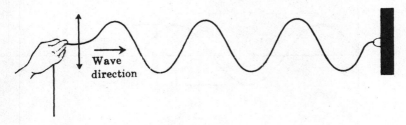

Figure 10-1

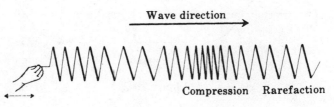

Figure 10-2

Wave Properties

The *period T* of a wave is the time required for one complete wave to pass a given point. The *frequency f* is the number of waves that pass that point per second [Figure 10-3(*a*)], so

$$f = \frac{1}{T}$$

$$\text{Frequency} = \frac{1}{\text{period}}$$

The *wavelength* λ (Greek letter *lambda*) of a periodic wave is the distance between adjacent wave crests (Figure 10-3(*b*)). Frequency and wavelength are related to wave velocity by

$$v = f\lambda$$

$$\text{Wave velocity} = (\text{frequency})\,(\text{wavelength})$$

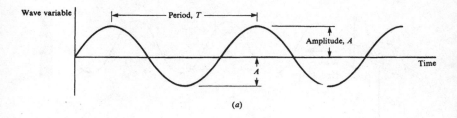

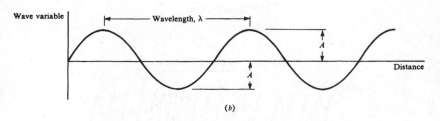

Figure 10-3

The *amplitude A* of a wave is the maximum displacement of the particles of the medium through which the wave passes on either side of their equilibrium position. In a transverse wave, the amplitude is half the distance between the top of a crest and the bottom of a trough (Figure 10-3).

The *intensity I* of a wave is the rate at which it transports energy per unit area perpendicular to the direction of motion. The intensity of a mechanical wave (one that involves moving matter, in contrast to, say, an electromagnetic wave) is proportional to f^2, the square of its frequency, and, to A^2 the square of its amplitude.

Solved Problem 10.1 The velocity of sound in seawater is 1531 m/s. Find the wavelength in seawater of a sound wave whose frequency is 256 Hz.

Solution. $\lambda = \dfrac{v}{f} = \dfrac{1531 \text{ m/s}}{256 \text{ Hz}} = 5.98 \text{ m}$

Logarithms

Although logarithms have many other uses, their chief application in applied physics is in connection with the decibel, which is described in the next section. Logarithms are discussed here only to the extent required for this purpose.

The *logarithm* of a number N is the power n to which 10 must be raised in order that $10^n = N$. That is,

$$N = 10^n \qquad \text{therefore} \qquad \log N = n$$

(Logarithms are not limited to a base of 10, but base-10 logarithms are the most common and are all that are needed here.) For instance,

$$1000 = 10^3 \qquad \text{therefore} \qquad \log 1000 = 3$$
$$0.01 = 10^{-2} \qquad \text{therefore} \qquad \log 0.01 = -2$$

To find the logarithm of a number with a calculator, enter the value of the number and press the LOG button.

The *antilogarithm* of a quantity n is the number N whose logarithm it is. That is,

$$\text{If} \qquad \log N = n \qquad \text{then} \qquad \text{antilog } n = N$$

To find the antilogarithm with a calculator, enter the value of the logarithm and press the INV LOG button.

Because of the way logarithms are defined, the logarithm of a product equals the sum of the logarithms of the factors:

$$\log xy = \log x + \log y$$

Other useful relations are

$$\log \frac{x}{y} = \log x - \log y$$

$$\log x^n = n \log x$$

Sound

Sound waves are longitudinal waves in which alternate regions of compression and rarefaction move away from a source. Sound waves can travel through solids, liquids, and gases. The velocity of sound is a constant for a given material at a giv-

en pressure and temperature; in air at 1-atm pressure and 20 °C, it is 343 m/s = 1125 ft/s.

When sound waves spread out uniformly in space, their intensity decreases inversely with the square of the distance R from their source. Thus, if the intensity of a certain sound is I_1 at the distance R_1, its intensity I_2 at the distance R_2 can be found from

$$\frac{I_2}{I_1} = \frac{R_1^2}{R_2^2}$$

The response of the human ear to sound intensity is not proportional to the intensity, so doubling the actual intensity of a certain sound does not lead to the sensation of a sound twice as loud but only of one that is slightly louder than the original. For this reason, the *decibel* (dB) scale is used for sound intensity.

An intensity of 10^{-12} W/m^2, which is just audible, is given the value 0 dB; a sound 10 times more intense is given the value 10 dB; a sound 10^2 times more intense than 0 dB is given the value of 20 dB; a sound 10^3 times more intense than 0 dB is given the value of 30 dB; and so forth. More formally, the intensity I dB of a sound wave whose intensity is I W/m^2 is given by

$$I \text{ dB} = 10 \log \frac{I}{I_o}$$

where $I_0 = 10^{-12}$ W/m^2. Normal conversation might be 60 dB, city traffic noise might be 90 dB, and a jet aircraft might produce as much as 140 dB (which produces damage to the ear) at a distance of 100 ft. Long-term exposure to intensity levels of over 85 dB usually leads to permanent hearing damage.

Solved Problem 10.2 How many times more intense is a 50-dB sound than a 40-dB sound? Than a 20-dB sound?

Solution. Each interval of 10 dB represents a change in sound intensity by a factor of 10. Hence a 50-dB sound is 10 times more intense than a 40-dB sound and $10 \times 10 \times 10 = 1000$ times more intense than a 20-dB sound.

Doppler Effect

When there is relative motion between a source of waves and an observer, the apparent frequency of the waves is different from their frequency f_S at the source. This change in frequency is called the *Doppler effect*. When the source approaches the observer (or vice versa), the observed frequency is higher; when the source recedes from the observer (or vice versa), the observed frequency is lower. In the case of sound waves, the frequency f that a listener hears is given by

$$f = f_S\left(\frac{v + v_L}{v - v_S}\right) \quad \text{sound}$$

In this formula, v is the velocity of sound, v_L is the velocity of the listener (considered positive for motion toward the source and negative for motion away from the source), and v_S is the velocity of the source (considered positive for motion toward the listener and negative for motion away from the listener).

The Doppler effect in electromagnetic waves (light and radio waves are examples) obeys the formula

$$f = f_S\left[\frac{1 + (v/c)}{1 - (v/c)}\right]^{1/2} \quad \text{electromagnetic waves}$$

Here c is the velocity of light (3.00×10^8 m/s), and v is the relative velocity between source and observer (considered positive if they are approaching and negative if they are receding).

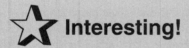 **Interesting!**

Astronomers use the Doppler effect in light to determine the motion of stars; police use the effect in radar waves to determine vehicle velocities.

Chapter 11
ELECTRICITY

Electric Charge

Electric charge, like mass, is one of the basic properties of certain elementary particles of which all matter is composed. There are two kinds of charge, *positive charge* and *negative charge*. The positive charge in ordinary matter is carried by *protons*, the negative charge by *electrons*. Charges of the same sign repel each other, charges of opposite sign attract each other.

The unit of charge is the *coulomb* (C). The charge of the proton is $+1.6 \times 10^{-19}$ C, and the charge of the electron is -1.6×10^{-19} C. All charges occur in multiples of $\pm e = \pm 1.6 \times 10^{-19}$ C.

According to the principle of *conservation of charge*, the net electric charge in an isolated system always remains constant. (Net charge means

70

the total positive charge minus the total negative charge.) When matter is created from energy, equal amounts of positive and negative charge always come into being, and when matter is converted to energy, equal amounts of positive and negative charge disappear.

Atoms and Ions

An *atom* of any element consists of a small, positively charged *nucleus* with a number of electrons some distance away. The nucleus is composed of protons (charge $+e$, mass $= 1.673 \times 10^{-27}$ kg) and neutrons (uncharged, mass $= 1.675 \times 10^{-27}$ kg). The number of protons in the nucleus is normally equal to the number of electrons around it, so the atom as a whole is electrically neutral. The forces between atoms that hold them together as solids and liquids are electric in origin. The mass of the electron is 9.1×10^{-31} kg.

Under certain circumstances, an atom may lose one or more electrons and become a *positive ion* or it may gain one or more electrons and become a *negative ion*. Many solids consist of positive and negative ions rather than of atoms or molecules. An example is ordinary table salt, which is made up of positive sodium ions (Na^+) and negative chlorine ions (Cl^-).

Coulomb's Law

The force one charge exerts on another is given by *Coulomb's law*:

$$\text{Electric force} \ = \ F = k\frac{q_1 q_2}{r^2}$$

where q_1 and q_2 are the magnitudes of the charges, r is the distance between them, and k is a constant whose value in free space is

$$k = 9.0 \times 10^9 \ \text{N·m}^2/\text{C}^2$$

The value of k in air is slightly greater. The constant k is sometimes replaced by

$$k = \frac{1}{4\pi\varepsilon_o}$$

where ε_0, the *permittivity of free space*, has the value

$$\varepsilon_0 = 8.85 \times 10^{-12} \ C^2/N \cdot m^2$$

(ε is the Greek letter *epsilon*.)

Solved Problem 11.1 Two charges, one of $+5 \times 10^{-7}$ C and the other -2×10^{-7} C, attract each other with a force of -100 N. How far apart are they?

Solution. From Coulomb's law, we have

$$r = \sqrt{\frac{kq_1q_2}{F}} = \sqrt{\frac{(9 \times 10^9 \ N \cdot m^2 / C^2)(5 \times 10^{-7} \ C)(-2 \times 10^{-7} \ C)}{-100 \ N}}$$
$$= \sqrt{90 \times 10^{-7} \ m^2} = 3 \text{ mm}$$

Electric Field

An *electric field* is a region of space in which a charge would be acted upon by an electric force. An electric field may be produced by one or more charges, and it may be uniform or it may vary in magnitude and/or direction from place to place.

If a charge q at a certain point is acted on by the force $\mathbf{F}$, the electric field $\mathbf{E}$ at that point is defined as the ratio between $\mathbf{F}$ and q:

$$\mathbf{E} = \frac{\mathbf{F}}{q}$$

$$\text{Electric field} = \frac{\text{force}}{\text{charge}}$$

Electric field is a vector quantity whose direction is that of the force on a positive charge. The unit of $\mathbf{E}$ is the newton per coulomb (N/C) or, more commonly, the equivalent unit volt per meter (V/m).

The advantage of knowing the electric field at some point is that we can at once establish the force on *any* charge q placed there, which is

$$\mathbf{F} = q\mathbf{E}$$

$$\text{Force} = (\text{charge}) \ (\text{electric field})$$

Electric Field Lines

Field lines are a means of describing a force field, such as an electric field, by using imaginary lines to indicate the direction and magnitude of the field. The direction of an electric field line at any point is the direction in which a positive charge would move if placed there, and field lines are drawn close together where the field is strong and far apart where the field is weak (Figure 11-1).

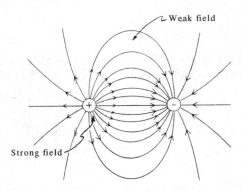

Figure 11-1

Solved Problem 11.2 The electric field in a certain neon sign is 5000 V/m. What force does this field exert on a neon ion of mass 3.3×10^{-26} kg and charge $+e$?

Solution. The force on the neon ion is

$$F = qE = eE = \left(1.6 \times 10^{-19}\ \text{C}\right)\left(5 \times 10^3\ \text{V}/\text{m}\right) = 8 \times 10^{-16}\ \text{N}$$

Potential Difference

The *potential difference* V between two points in an electric field is the amount of work needed to take a charge of 1 C from one of the points to the other. Thus

$$V = \frac{W}{q}$$

$$\text{Potential difference} = \frac{\text{work}}{\text{charge}}$$

The unit of potential difference is the *volt* (V):

$$1 \text{ volt} = 1 \ \frac{\text{joule}}{\text{coulomb}}$$

The potential difference between two points in a uniform electric field **E** is equal to the product of E and the distance s between the points in a direction parallel to **E**:

$$V = Es$$

Since an electric field is usually produced by applying a potential difference between two metal plates s apart, this equation is most useful in the form

$$E = \frac{V}{s}$$

$$\text{Electric field} = \frac{\text{potential difference}}{\text{distance}}$$

You Need to Know ✔

A battery uses chemical reactions to produce a potential difference between its terminals; a generator uses electromagnetic induction for this purpose.

Solved Problem 11.3 The potential difference between a certain thundercloud and the ground is 7×10^6 V. Find the energy dissipated when a charge of 50 C is transferred from the cloud to the ground in a lightning stroke.

Solution. The energy is

$$W = qV = (50 \text{ C})(7 \times 10^6 \text{ V}) = 3.5 \times 10^8 \text{ J}$$

Chapter 12
ELECTRIC CURRENT

Electric Current

A flow of charge from one place to another constitutes an *electric current*. An *electric circuit* is a closed path in which an electric current carries energy from a source (such as a battery or generator), to a load (such as a motor or a lamp). In such a circuit (see Figure 12-1), electric current

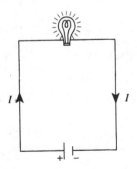

Figure 12-1

is assumed to go from the positive terminal of the battery (or generator) through the circuit and back to the negative terminal of the battery. The direction of the current is conventionally considered to be that in which a positive charge would have to move to produce the same effects as the actual current. Thus a current is always supposed to go from the positive terminal of a battery to its negative terminal.

A *conductor* is a substance through which charge can flow easily, and an *insulator* is one through which charge can flow only with great difficulty. Metals, many liquids, and *plasmas* (gases whose molecules are charged) are conductors; nonmetallic solids, certain liquids, and gases whose molecules are electrically neutral are insulators.

 **Note!**

A number of substances, called *semiconductors*, are intermediate in their ability to conduct charge.

Electric currents in metal wires always consist of flows of electrons; such currents are assumed to occur in the direction opposite to that in which the electrons move. Since a positive charge going one way is, for most purposes, equivalent to a negative charge going the other way, this assumption makes no practical difference. Both positive and negative charges move when a current is present in a liquid or gaseous conductor.

If an amount of charge q passes a given point in a conductor in the time interval t, the current in the conductor is

$$I = \frac{q}{t}$$

$$\text{Electric current} = \frac{\text{charge}}{\text{time interval}}$$

The unit of electric current is the *ampere* (A), where

$$1 \text{ ampere} = 1 \frac{\text{coulomb}}{\text{second}}$$

Ohm's Law

For a current to exist in a conductor, there must be a
potential difference between its ends, just as a difference in height between source and outlet is necessary for a river current to exist. In the case of a metal-
lic conductor, the current is proportional to the
applied potential difference: Doubling V causes I to double, tripling V
causes I to triple, and so forth. This relationship is known as *Ohm's law*
and is expressed in the form

$$I = \frac{V}{R}$$

$$\text{Electric current} = \frac{\text{potential difference}}{\text{resistance}}$$

The quantity R is a constant for a given conductor and is called its *resistance*. The unit of resistance is the *ohm* (Ω), where

$$1 \text{ ohm} = 1 \frac{\text{volt}}{\text{ampere}}$$

The greater the resistance of a conductor, the less the current when a certain potential difference is applied.

Ohm's law is not a physical principle but is an experimental relationship that most metals obey over a wide range of values of V and I.

Solved Problem 12.1 A 120-V electric heater draws a current of 25 A.
What is its resistance?

Solution.

$$R = \frac{V}{I} = \frac{120 \text{ V}}{25 \text{ A}} = 4.8 \text{ A}$$

Resistivity

The resistance of a conductor that obeys Ohm's law is given by

$$R = \rho \frac{L}{A}$$

where L is the length of the conductor, A is the cross-sectional area, and ρ (Greek letter *rho*), is the *resistivity* of the material of the conductor. In SI, the unit of resistivity is the ohm-meter.

The resistivities of most materials vary with temperature. If R is the resistance of a conductor at a particular temperature, then the change in its resistance ΔR when the temperature changes by ΔT is approximately proportional to both R and ΔT so that

$$\Delta R = \alpha R \Delta T$$

The quantity α is the *temperature coefficient of resistance* of the material.

Electric Power

The rate at which work is done to maintain an electric current is given by the product of the current I and the potential difference V:

$$P = IV$$

Power = (current) (potential difference)

When I is in amperes and V is in volts, P will be in watts.

If the conductor or device through which a current passes obeys Ohm's law, the power consumed may be expressed in the alternative forms

$$P = IV = I^2 R = \frac{V^2}{R}$$

Table 12.1 is a summary of the various formulas for potential difference V, current I, resistance R, and power P that follow from Ohm's law $I = V/R$ and from the power formula $P = VI$.

Solved Problem 12.2 The current through a 50-Ω resistance is 2 A. How much power is dissipated as heat?

Solution. $P = I^2 R = (2 \text{ A})^2 (50 \text{ Ω}) = 200 \text{ W}$

Unknown Quantity	Known Quantities					
	V, I	I, R	V, R	P, I	P, V	P, R
$V =$		IR		P/I		$\sqrt{PR}$
$I =$			V/R		P/V	$\sqrt{P/R}$
$R =$	V/I			P/I^2	V^2/P	
$P =$	VI	I^2R	V^2/R			

Table 12.1

Chapter 13
DIRECT-CURRENT CIRCUITS

IN THIS CHAPTER:

✔ *Resistors in Series*
✔ *Resistors in Parallel*
✔ *EMF and Internal Resistance*
✔ *Kirchhoff's Rules*

Resistors in Series

The equivalent resistance of a set of resistors depends on the way in which they are connected as well as on their values. If the resistors are joined in *series*, that is, consecutively (Figure 13-1), the equivalent resistance R of the combination is the sum of the individual resistances:

$$R = R_1 + R_2 + R_3 + \cdots \quad \text{series resistors}$$

Resistors in Parallel

In a *parallel* set of resistors, the corresponding terminals of the resistors are connected (Figure 13-2). The reciprocal $1/R$ of the equivalent resistance of the combination is the sum of the reciprocals of the individual resistances:

$$\frac{1}{R} = \frac{1}{R_1} + \frac{1}{R_2} + \frac{1}{R_3} + \cdots \quad \text{parallel resistors}$$

If only two resistors are connected in parallel,

$$\frac{1}{R} = \frac{1}{R_1} + \frac{1}{R_2} = \frac{R_1 + R_2}{R_1 R_2} \quad \text{and so} \quad R = \frac{R_1 R_2}{R_1 + R_2}$$

Solved Problem 13.1 Find the equivalent resistance of the circuit shown in Figure 13-3(a).

Solution. Figure 13-3(b) shows how the original circuit is decomposed into its series and parallel parts, each of which is treated in turn. The equivalent resistance of R_1 and R_2 is

$$R' = \frac{R_1 R_2}{R_1 + R_2} = \frac{(10 \ \Omega)(10 \ \Omega)}{10 \ \Omega + 10 \ \Omega} = 5 \ \Omega$$

This equivalent resistance is in series with R_3, and so

$$R'' = R' + R_3 = 5 \ \Omega + 3 \ \Omega = 8 \ \Omega$$

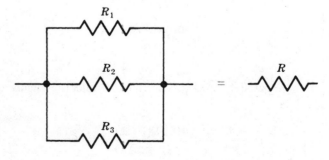

Figure 13-1

Figure 13-2

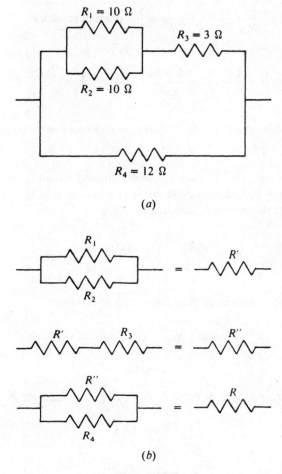

(a)

(b)

Figure 13-3

Finally R'', is in parallel with R_4; hence the equivalent resistance of the entire circuit is

$$R = \frac{R''R_4}{R'' + R_4} = \frac{(8\ \Omega)(12\ \Omega)}{8\ \Omega + 12\ \Omega} = 4.8\ \Omega$$

EMF and Internal Resistance

The work done per coulomb on the charge passing through a battery, generator, or other source of electric energy is called the *electromotive force*, or *emf*, of the source. The emf is equal to the potential difference across the terminals of the source when no current flows. When a current I flows, this potential difference is less than the emf because of the *internal resistance* of the source. If the internal resistance is r, then a potential drop of Ir occurs within the source. The terminal voltage V across a source of emf V_e whose internal resistance is r when it provides a current of I is therefore

$$V = V_e - Ir$$

Terminal voltage = emf − potential drop due to internal resistance

When a battery or generator of emf V_e is connected to an external resistance R, the total resistance in the circuit is $R + r$, and the current that flows is

$$I = \frac{V_e}{R + r}$$

$$\text{Current} = \frac{\text{emf}}{\text{external resistance} + \text{internal resistance}}$$

Kirchhoff's Rules

The current that flows in each branch of a complex circuit can be found by applying *Kirchhoff's rules* to the circuit. The first rule applies to *junctions* of three or more wires (Figure 13-4) and is a consequence of conservation of charge. The second rule applies to *loops*, which are closed conducting paths in the circuit, and is a consequence of conservation of energy. The rules are:

1. The sum of the currents that flow into a junction is equal to the sum of the currents that flow out of the junction.
2. The sum of the emfs around a loop is equal to the sum of the *IR* potential drops around the loop.

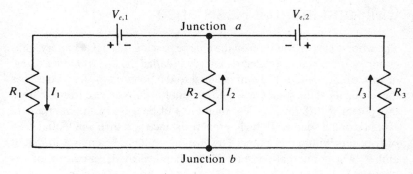

Figure 13-4

The procedure for applying Kirchhoff's rules is as follows:

1. Choose a direction for the current in each branch of the circuit, as in Figure 13-4. (A *branch* is a section of a circuit between two junctions.) If the choice is correct, the current will turn out to be positive. If not, the current will turn out to be negative, which means that the actual current is in the opposite direction. The current is the same in all the resistors and emf sources in a given branch. Of course, the currents will usually be different in the different branches.

2. Apply the first rule to the currents at the various junctions. This gives as many equations as the number of junctions. However, one of these equations is always a combination of the others and so gives no new information. (If there are only two junction equations, they will be the same.) Thus, the number of usable junction equations is equal to one less than the number of junctions.

3. Apply the second rule to the emfs and IR drops in the loops. In going around a loop (which can be done either clockwise or counterclockwise), an emf is considered positive if the negative terminal of its source is met first. If the positive terminal is met first, the emf is considered negative. An IR drop is considered positive if the current in the resistor R is in the same direction as the path being followed. If the current direction is opposite to the path, the IR drop is considered negative.

In the case of the circuit shown in Figure 13-4, Kirchhoff's first rule, applied to either junction a or junction b, yields

$$I_1 = I_2 + I_3$$

The second rule applied to loop 1, shown in Figure 13-5(a), and proceeding counterclockwise, yields

$$V_{e,1} = I_1 R_1 + I_2 R_2$$

The rule applied to loop 2, shown in Figure 13-5(b), and again proceeding counterclockwise, yields

$$-V_{e,2} = -I_2 R_2 + I_3 R_3$$

There is also a third loop, namely, the outside one shown in Figure 13-5(c), which must similarly obey Kirchhoff's second rule. For the sake of variety, we now proceed clockwise and obtain

$$-V_{e,1} + V_{e,2} = -I_3 R_3 - I_1 R_1$$

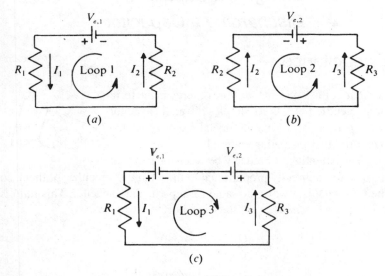

(a) (b)

(c)

Figure 13-5

Chapter 14
CAPACITANCE

Capacitance

A *capacitor* is a system that stores energy in the form of an electric field. In its simplest form, a capacitor consists of a pair of parallel metal plates separated by air or other insulating material.

The potential difference V between the plates of a capacitor is directly proportional to the charge Q on either of them, so the ratio Q/V is always the same for a particular capacitor. This ratio is called the *capacitance* C of the capacitor:

$$C = \frac{Q}{V}$$

$$\text{Capacitance} = \frac{\text{charge on either plate}}{\text{potential difference between plates}}$$

The unit of capacitance is the *farad* (F), where 1 farad = 1 coulomb/ volt. Since the farad is too large for practical purposes, the *microfarad* and *picofarad* are commonly used, where

$$1 \text{ microfarad} = 1\,\mu\text{F} = 10^{-6} \text{ F}$$
$$1 \text{ picofarad} = 1\,p\text{F} = 10^{-12} \text{ F}$$

A charge of 10^{-6} C on each plate of 1-μF capacitor will produce a potential difference of $V = Q/C = 1$ V between the plates.

Parallel-Plate Capacitor

A capacitor that consists of parallel plates each of area A separated by the distance d has a capacitance of

$$C = K\varepsilon_o \frac{A}{d}$$

The constant ε_0 is the permittivity of free space; its value is

$$\varepsilon_o = 8.85 \times 10^{-12} \text{ C}^2 / \left(\text{N} \cdot \text{m}^2\right) = 8.85 \times 10^{-12} \text{ F/m}$$

The quantity K is the *dielectric constant* of the material between the capacitor plates; the greater K is, the more effective the material is in diminishing an electric field.

 **Note!**

For free space, $K = 1$; for air, $K = 1.0006$; a typical value for glass is $K = 6$; and for water, $K = 80$.

Capacitors in Combination

The *equivalent capacitance* of a set of connected capacitors is the capacitance of the single capacitor that can replace the set without changing the

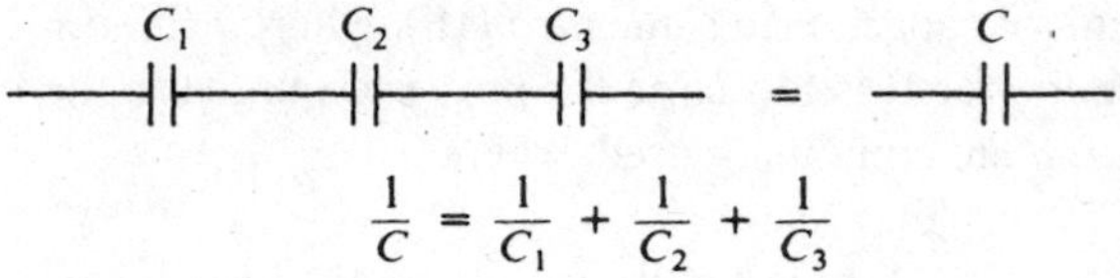

$$\frac{1}{C} = \frac{1}{C_1} + \frac{1}{C_2} + \frac{1}{C_3}$$

Figure 14-1

properties of any circuit it is part of. The equivalent capacitance of a set of capacitors joined in series (Figure 14-1) is

$$\frac{1}{C} = \frac{1}{C_1} + \frac{1}{C_2} + \frac{1}{C_3} + \cdots \quad \text{capacitors in series}$$

If there are only two capacitors in series,

$$\frac{1}{C} = \frac{1}{C_1} + \frac{1}{C_2} = \frac{C_1 + C_2}{C_1 C_2} \quad \text{and so} \quad C = \frac{C_1 C_2}{C_1 + C_2}$$

In a parallel set of capacitors (Figure 14-2),

$$C = C_1 + C_2 + C_3 + \cdots \quad \text{capacitors in parallel}$$

Solved Problem 14.1 Find the equivalent capacitance of three capacitors whose capacitances are 1, 2, and 3 μF that are connected in: (*a*) series and (*b*) parallel.

Solution.

(*a*) In series, the equivalent capacitance can be found by:

$$\frac{1}{C} = \frac{1}{C_1} + \frac{1}{C_2} + \frac{1}{C_3} = \frac{1}{1\,\mu F} + \frac{1}{2\,\mu F} + \frac{1}{3\,\mu F} = \frac{11}{6\,\mu F}$$

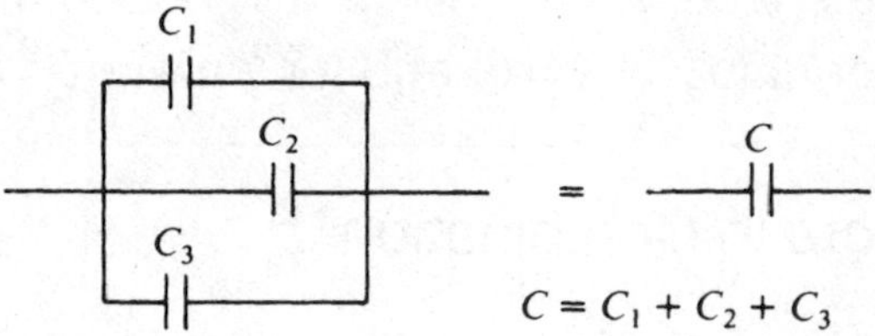

$$C = C_1 + C_2 + C_3$$

Figure 14-2

$$C = \frac{6}{11} \ \mu F = 0.545 \ \mu F$$

(*b*) In parallel, the equivalent capacitance can be found by:

$$C = C_1 + C_2 + C_3 = 1 \ \mu F + 2 \ \mu F + 3 \ \mu F = 6 \ \mu F$$

Energy of a Charged Capacitor

To produce the electric field in a charged capacitor, work must be done to separate the positive and negative charges. This work is stored as electric potential energy in the capacitor. The potential energy W of a capacitor of capacitance C whose charge is Q and whose potential difference is V given by

$$W = \frac{1}{2}QV = \frac{1}{2}CV^2 = \frac{1}{2}\left(\frac{Q^2}{C}\right)$$

Charging a Capacitor

When a capacitor is being charged in a circuit such as that of Figure 14-3, at any moment the voltage Q/C across it is in the opposite direction to the battery voltage V and thus tends to oppose the flow of additional charge. For this reason, a capacitor does not acquire its final charge the instant it is connected to a battery or other source of emf. The net potential difference when the charge on the capacitor is Q is $V - (Q/C)$, and the current is then

$$I = \frac{\Delta Q}{\Delta t} = \frac{V - (Q/C)}{R}$$

As Q increases, its rate of increase $I = \Delta Q/\Delta t$ decreases. Figure 14-4 shows how Q, measured in percent of final charge, varies with time when a capacitor is being charged; the switch of Figure 14-3 is closed at $t = 0$.

The product RC of the resistance R in the circuit and the capacitance C governs the rate at which the capacitor reaches its ultimate charge of $Q_0 = CV$. The product RC is called the *time constant T* of the circuit. Af-

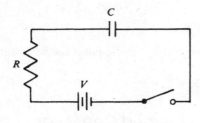

Figure 14-3

ter a time equal to T, the charge on the capacitor is 63 percent of its final value.

The formula that governs the growth of charge in the circuit of Figure 14-3 is

$$Q = Q_0\left(1 - e^{-t/T}\right)$$

where Q_o is the final charge CV and T is the time constant RC. Figure 14-4 is a graph of that formula. It is easy to see why Q reaches 63 percent of Q_o in time T. When $t = T$, $t/T = 1$ and

$$Q = Q_0\left(1 - e^{-t/T}\right) = Q_0\left(1 - e^{-1}\right) = Q_0\left(1 - \frac{1}{e}\right)$$

$$= Q_0(1 - 0.37) = 0.63Q_0$$

Discharging a Capacitor

When a charged capacitor is discharged through a resistance, as in Figure 14-5, the decrease in charge is governed by the formula

$$Q = Q_0 e^{-t/T}$$

where again $T = RC$ is the time constant. The charge will fall to 37 percent of its original value after time T (Figure 14-6). The smaller the time constant T, the more rapidly a capacitor can be charged or discharged.

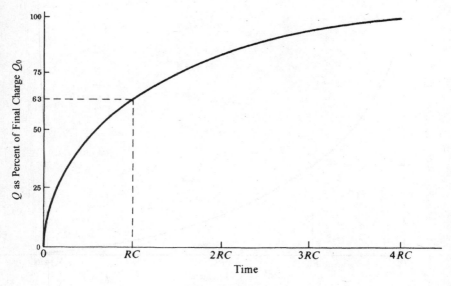

Figure 14-4

Solved Problem 14.2 A 20-μF capacitor is connected to a 45-V battery through a circuit whose resistance is 2000 Ω. (*a*) What is the final charge on the capacitor? (*b*) How long does it take for the charge to reach 63 percent of its final value?

Solution. (*a*)
$$Q = CV = (20 \times 10^{-6} \text{ F})(45 \text{ V}) = 9 \times 10^{-4} \text{ C}$$

(*b*)
$$t = RC = (2000 \text{ Ω})(20 \times 10^{-6} \text{ F}) = 0.04 \text{ s}$$

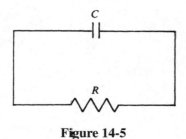

Figure 14-5

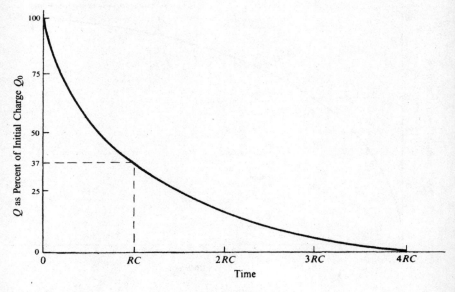

Figure 14-6

Chapter 15
MAGNETISM

Nature of Magnetism

Two electric charges at rest exert forces on each other according to Coulomb's law. When the charges are in motion, the forces are different, and it is customary to attribute the differences to *magnetic forces* that occur between moving charges in addition to the electric forces between them. In this interpretation, the total

force on a charge Q at a certain time and place can be divided into two parts: an electric force that depends only on the value of Q and a magnetic force that depends on the velocity v of the charge as well as on Q.

In reality, there is only a single interaction between charges, the *electromagnetic interaction*. The theory of relativity provides the link between electric and magnetic forces: Just as the mass of an object moving with respect to an observer is greater than when it is at rest, so the electric force between two charges appears altered to an observer when the charges are moving with respect to the observer. Magnetism is not distinct from electricity in the way that, for example, gravitation is.

You Need to Know

Despite the unity of the electromagnetic interaction, it is convenient for many purposes to treat electric and magnetic effects separately.

Magnetic Field

A *magnetic field* **B** is present wherever a magnetic force acts on a moving charge. The direction of **B** at a certain place is that along which a charge can move without experiencing a magnetic force; along any other direction that the charge would be acted on by such a force. The magnitude of **B** is equal numerically to the force on a charge of 1 C moving at 1 m/s perpendicular to **B**.

The unit of magnetic field is the *tesla* (T), where

$$1 \text{ tesla} = 1 \frac{\text{newton}}{\text{ampere-meter}} = 1 \frac{\text{weber}}{(\text{meter})^2}$$

The *gauss* (G), equal to 10^{-4} T, is another unit of magnetic field sometimes used.

Magnetic Field of a Straight Current

The magnetic field a distance s from a long, straight current I has the magnitude

$$B = \left(\frac{\mu}{2\pi}\right)\left(\frac{1}{s}\right) \quad \text{straight current}$$

where μ is the *permeability* of the medium in which the magnetic field exists. The permeability of free space μ_0 has the value

$$\mu_0 = 4\pi \times 10^{-7} \text{ T·m/A} = 1.257 \times 10^{-6} \text{ T·m/A}$$

The field lines of the magnetic field around a straight current are in the form of concentric circles around the current. To find the direction of **B**, place the thumb of the right hand in the direction of the current; the curled fingers of that hand then point in the direction of **B** (Figure 15-1).

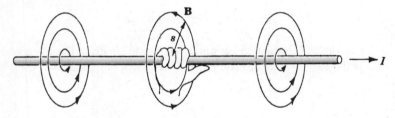

Figure 15-1

Magnetic Field of a Current Loop

The magnetic field at the center of a current loop of radius r has the magnitude

$$B = \left(\frac{\mu}{2}\right)\left(\frac{I}{r}\right) \quad \text{current loop}$$

The field lines of **B** are perpendicular to the plane of the loop, as shown in Figure 15-2(*a*). To find the direction of **B**, grasp the loop so the curled fingers of the right hand point in the direction of the current; the thumb of that hand then points in the direction of **B** [Figure 15-2(*b*)].

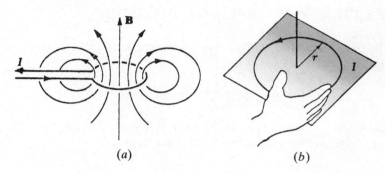

(a)

(b)

Figure 15-2

A *solenoid* is a coil consisting of many loops of wire. If the turns are close together and the solenoid is long compared with its diameter, the magnetic field inside it is uniform and parallel to the axis with magnitude

$$B = \mu \frac{N}{L} I \quad \text{separate solenoid}$$

In this formula, N is the number of turns, L is the length of the solenoid, and I is the current. The direction of **B** is as shown in Figure 15-3.

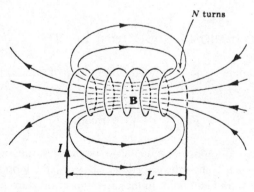

Figure 15-3

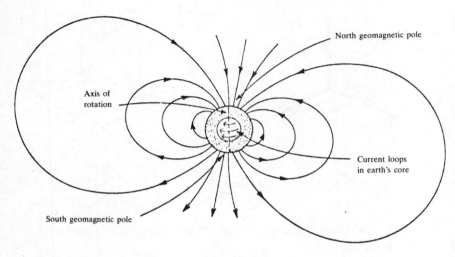

Figure 15-4

Earth's Magnetic Field

The earth has a magnetic field that arises from electric currents in its liquid iron core. The field is like that which would be produced by a current
loop centered a few hundred miles from the earth's center whose plane is
tilted by 11° from the plane of the equator (Figure 15-4). The *geomag-
netic poles* are the points where the magnetic axis passes through the
earth's surface. The magnitude of the earth's magnetic field varies from
place to place; a typical sea-level value is

$$3 \times 10^{-5} \text{ T.}$$

Solved Problem 15.1 In what ways are electric and magnetic fields sim-
ilar? In what ways are they different?

Solution.

Similarities: Both fields originate in electric charges, and both fields
can exert forces on electric charges.

Differences: All electric charges give rise to electric fields, but only a
charge in motion relative to an observer gives rise to a magnetic field.

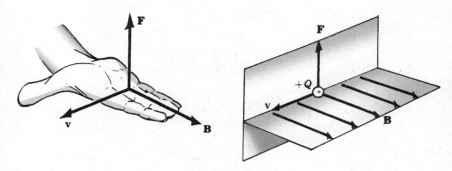

Figure 15-5

Electric fields exert forces on all charges, but magnetic fields exert forces only on moving charges.

Magnetic Force on a Moving Charge

The magnetic force on a moving charge Q in a magnetic field varies with the relative directions of $\mathbf{v}$ and $\mathbf{B}$. When $\mathbf{v}$ is parallel to $\mathbf{B}$, $F = 0$; when $\mathbf{v}$ is perpendicular to $\mathbf{B}$, F has its maximum value of

$$F = QvB \qquad \mathbf{v} \perp \mathbf{B}$$

The direction of $\mathbf{F}$ in the case of a positive charge is given by the right-hand rule, shown in Figure 15-5; $\mathbf{F}$ is in the opposite direction when the charge is negative.

Magnetic Force on a Current

Since a current consists of moving charges, a current-carrying wire will experience no force when parallel to a magnetic field $\mathbf{B}$ and maximum force when perpendicular to $\mathbf{B}$. In the latter case, F has the value

$$F = ILB \qquad \mathbf{I} \perp \mathbf{B}$$

where I is the current and L is the length of wire in the magnetic field. The direction of the force is as shown in Figure 15-6.

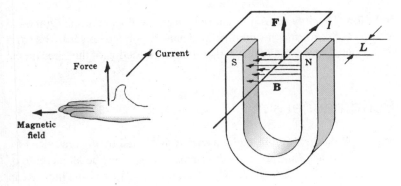

Figure 15-6

⭐ Note!

Owing to the different forces exerted on each of its sides, a current loop in a magnetic field always tends to rotate so that its plane is perpendicular to **B**. This effect underlies the operation of all electric motors.

Force Between Two Currents

Two parallel electric currents exert magnetic forces on each other (Figure 15-7). If the currents are in the same direction, the forces are attractive; if the currents are in opposite directions, the forces are repulsive. The force per unit length F/L on each current depends on currents I_1 and I_2 and their separation s:

$$\frac{F}{L} = \left(\frac{\mu_o}{2\pi}\right)\left(\frac{I_1 I_2}{s}\right) \quad \text{parallel currents}$$

Solved Problem 15.2 A positive charge is moving virtually upward when it enters a magnetic field directed to the north. In what direction is the force on the charge?

Solution. To apply the right-hand rule here, the fingers of the right hand are pointed north and the thumb of that hand is pointed upward. The palm of the hand faces west, which is therefore the direction of the force on the charge.

Ferromagnetism

The magnetic field produced by a current is altered by the presence of a substance of any kind. Usually the change, which may be an increase or a decrease in **B**, is very small, but in certain cases, there is an increase in **B** by hundreds or thousands of times. Substances that have the latter effect are called *ferromagnetic*; iron and iron alloys are familiar examples.

Remember

An *electromagnet* is a solenoid with a ferromagnetic core to increase its magnetic field.

Ferromagnetism is a consequence of the magnetic properties of the electrons that all atoms contain. An electron behaves in some respects as though it is a spinning charged sphere, and it is therefore magnetically equivalent to a tiny current loop. In most substances, the magnetic fields of the atomic electrons cancel, but in ferromagnetic substances, the cancellation is not complete and each atom has a certain magnetic field of its own. The atomic magnetic fields align themselves in groups called *domains* with an external magnetic field to produce a much stronger total **B**. When the external field is removed, the atomic magnetic fields may remain aligned to produce a *permanent magnet*. The field of a bar magnet has the same form as that of a solenoid because both fields are due to parallel current loops (Figure 15-7).

Magnetic Intensity

A substance that decreases the magnetic field of a current is called *diamagnetic*; it has a permeability μ that is less than μ_0. Copper and water

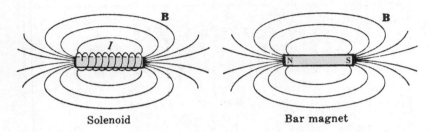

Solenoid   Bar magnet

Figure 15-7

are examples. A substance that increases the magnetic field of a current by a small amount is called *paramagnetic*; it has a permeability μ that is greater than μ_0. Aluminum is an example. Ferromagnetic substances have permeabilities hundreds or thousands of times greater than μ_0.

⭐ **Note!**

Diamagnetic substances are repelled by magnets; paramagnetic and ferromagnetic ones are attracted by magnets.

Because different substances have different magnetic properties, it is useful to define a quantity called *magnetic intensity* **H**, which is independent of the medium in which a magnetic field is located. The magnetic intensity in a place where the magnetic field is **B** and the permeability is μ is given by

$$\mathbf{H} = \frac{\mathbf{B}}{\mu}$$

$$\text{Magnetic intensity} = \frac{\text{magnetic field}}{\text{permeability of medium}}$$

The unit of **H** is the ampere per meter. Magnetic intensity is sometimes called *magnetizing force* or *magnetizing field*.

The permeability of a ferromagnetic material at a given value of H varies both with H and with the previous degree of magnetization of the material. The latter effect is known as *hysteresis*.

Chapter 16
ELECTRO-MAGNETIC INDUCTION

IN THIS CHAPTER:

- ✔ *Electromagnetic Induction*
- ✔ *Faraday's Law*
- ✔ *Lenz's Law*
- ✔ *The Transformer*
- ✔ *Self-Inductance*
- ✔ *Inductors in Combination*
- ✔ *Energy of a Current-Carrying Inductor*

Electromagnetic Induction

A current is produced in a conductor whenever the current cuts across magnetic field lines, a phenomenon known as *electromagnetic induction*. If the motion is parallel to the field lines of force, there is no effect. Electromagnetic induction originates in the force a magnetic field exerts on a moving charge. When a wire moves across a magnetic field, the electrons it contains

experience sideways forces that push them along the wire to cause a current. It is not even necessary for there to be relative motion of a wire and a source of magnetic field, since a magnetic field whose strength is changing has moving field lines associated with it and a current will be induced in a conductor that is in the path of these moving field lines.

When a straight conductor of length l is moving across a magnetic field **B** with the velocity **v**, the emf induced in the conductor is given by

$$\text{Induced emf} = V_e = Blv$$

when **B**, **v**, and the conductor are all perpendicular to one another.

Solved Problem 16.1 The vertical component of the earth's magnetic field in a certain region is 3×10^{-5} T. What is the potential difference between the rear wheels of a car, which are 1.5 m apart, when the car's velocity is 20 m/s?

Solution. The real axle of the car may be considered as a rod of 1.5 m long-moving perpendicular to the magnetic field's vertical component. The potential difference between the wheels is therefore

$$V_e = Blv = (3 \times 10^{-5} \text{ T})(1.5\text{m})(20 \text{ m/s}) = 9 \times 10^{-4} \text{ V} = 0.9 \text{ mV}$$

Faraday's Law

Figure 16-1 shows a coil (called a *solenoid*) of N turns that encloses an area A. The axis of the coil is parallel to a magnetic field **B**. According to *Faraday's law of electromagnetic induction*, the emf induced in the coil when the product BA changes by $\Delta(BA)$ in the time Δt is given by

$$\text{Induced emf} = V_e = -N\frac{\Delta(BA)}{\Delta t}$$

The quantity BA is called the *magnetic flux* enclosed by the coil and is denoted by the symbol Φ (Greek capital letter *phi*):

$$\Phi = BA$$

Magnetic flux = (magnetic field) (cross-sectional area)

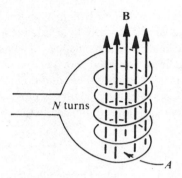

Figure 16-1

The unit of magnetic flux is the *weber* (Wb), where $1 \text{ Wb} = 1 \text{ T·m}^2$. Thus, Faraday's law can be written

$$V_e = -N \frac{\Delta \Phi}{\Delta t}$$

Lenz's Law

The minus sign in Faraday's law is a consequence of *Lenz's law*:
An induced current is always in such a direction that its own magnetic field acts to oppose the effect that created it.

For example, if **B** is decreasing in magnitude in the situation of Figure 16-1, the induced current in the coil will be counterclockwise in order that its own magnetic field will add to **B** and so reduce the rate at which **B** is decreasing. Similarly, if **B** is increasing, the induced current in the coil will be clockwise so that its own magnetic field will subtract from **B** and thus reduce the rate at which **B** is increasing.

The Transformer

A *transformer* consists of two coils of wire, usually wound on an iron core. When an alternating current is passed through one of the windings, the changing magnetic field it gives rise to induces an alternating current in the other winding. The potential difference per turn is the same in both primary and secondary windings, so the ratio of turns in the winding determines the ratio of voltages across them:

$$\frac{V_1}{V_2} = \frac{N_1}{N_2}$$

$$\frac{\text{Primary voltage}}{\text{Secondary voltage}} = \frac{\text{Primary turns}}{\text{Secondary turns}}$$

Since the power $I_1 V_1$ going into a transformer must equal the power $I_2 V_2$ going out, where I_1 and I_2 are the primary and secondary currents, respectively, the ratio of currents is inversely proportional to the ratio of turns:

$$\frac{I_1}{I_2} = \frac{N_2}{N_1}$$

$$\frac{\text{Primary current}}{\text{Secondary current}} = \frac{\text{Secondary turns}}{\text{Primary turns}}$$

Self-Inductance

When the current in a circuit changes, the magnetic field enclosed by the circuit also changes, and the resulting change in flux leads to a *self-induced emf* of

$$\text{Self-induced emf} = V_e = -L\frac{\Delta I}{\Delta t}$$

Here $\Delta I / \Delta t$ is the rate of change of the current, and L is a property of the circuit called its *self-inductance*, or, more commonly, its *inductance*. The minus sign indicates that the direction of V_e is such as to oppose the change in current ΔI that caused it.

The unit of inductance is the *henry* (H). A circuit or circuit element that has an inductance of 1 H will have a self-induced emf of 1 V when the current through it changes at the rate of 1 A/s. Because the henry is a rather large unit, the *millihenry* and *microhenry* are often used, where

$$1 \text{ millihenry} = 1 \text{ mH} = 10^{-3} \text{ H}$$
$$1 \text{ microhenry} = 1 \text{ } \mu\text{H} = 10^{-6} \text{ H}$$

A circuit element with inductance is called an *inductor*. A solenoid is an example of an inductor. The inductance of a solenoid is

$$L = \frac{\mu N^2 A}{l}$$

where μ is the permeability of the core material, N is the number of turns, A is the cross-sectional area, and l is the length of the solenoid.

Inductors in Combination

When two or more inductors are sufficiently far apart for them not to interact electromagnetically, their equivalent inductances when they are connected in series and in parallel are as follows:

$$L = L_1 + L_2 + L_3 + \cdots \quad \text{inductors in series}$$

$$\frac{1}{L} = \frac{1}{L_1} + \frac{1}{L_2} + \frac{1}{L_3} + \cdots \quad \text{inductors in parallel}$$

Connecting coils in parallel reduces the total inductance to less than that of any of the individual coils. This is rarely done because coils are relatively large and expensive compared with other electronic components; a coil of the required smaller inductance would normally be used in the first place.

Because the magnetic field of a current-carrying coil extends beyond the inductor itself, the total inductance of two or more connected coils will be changed if they are close to one another. Depending on how the coils are arranged, the total inductance may be larger or smaller than if the coils were farther apart. This effect is called *mutual inductance* and is not considered in the above formula.

Solved Problem 16.2 Find the equivalent inductances of a 5- and an 8-mH inductor when they are connected in (*a*) series and (*b*) parallel.

Solution.

(*a*) $$L = L_1 + L_2 = 5 \text{ mH} + 8 \text{ mH} = 13 \text{ mH}$$

(b) $$\frac{1}{L} = \frac{1}{L_1} + \frac{1}{L_2} = \frac{1}{5 \text{ mH}} + \frac{1}{8 \text{ mH}} \qquad L = 3.08 \text{ mH}$$

Energy of a Current-Carrying Inductor

Because a self-induced emf opposes any change in an inductor, work has to be done against this emf to establish a current in the inductor. This work is stored as magnetic potential energy. If L is the inductance of an inductor, its potential energy when it carries the current I is

$$W = \frac{1}{2} L I^2$$

This energy powers the self-induced emf that opposes any decrease in the current through the inductor.

Chapter 17
LIGHT

Electromagnetic Waves

Electromagnetic waves consist of coupled electric and magnetic fields that vary periodically as they move through space. The electric and magnetic fields are perpendicular to each other and to the direction in which the waves travel (Figure 17-1), so the waves are transverse, and the variations in **E** and **B** occur simultaneously. Electromagnetic waves transport energy and require no material medium for their passage. Radio waves, light waves, X-rays, and gamma rays are examples of electromagnetic waves, and they differ only in frequency.

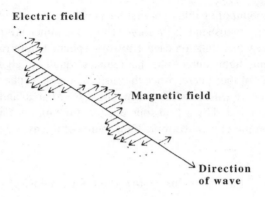

Figure 17-1

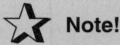 **Note!**

The color sensation produced by light waves depends on their frequency, with red light having the lowest visible frequencies and violet light the highest. White light contains light waves of all visible frequencies.

Electromagnetic waves are generated by accelerated electric charges, usually electrons. Electrons oscillating back and forth in an antenna give off radio waves, for instance, and accelerated electrons in atoms give off light waves.

In free space, all electromagnetic waves have the *velocity of light* which is

$$\text{Velocity of light} = c = 3.00 \times 10^8 \text{ m/s} = 186{,}000 \text{ mi/s}$$

Luminous Intensity and Flux

The brightness of a light source is called its *luminous intensity I*, whose unit is the *candela* (cd).The intensity of a light source is sometimes referred to as its *candlepower*.

The amount of visible light that falls on a given surface is called *luminous flux F*, whose unit is the *lumen* (lm). One lumen is equal to the luminous flux which falls on each 1 m^2 of a sphere 1 m in radius when a 1-cd isotropic light source (one that radiates equally in all directions) is at the center of the sphere. Since the surface area of a sphere of radius r is $4\pi r^2$, a sphere whose radius is 1 m has 4π m^2 of area, and the total luminous flux emitted by a 1-cd source is therefore 4π lm. Thus the luminous flux emitted by an isotropic light source of intensity I is given by

$$F = 4\pi I$$

Luminous flux $= (4\pi)(\text{luminous intensity})$

The above formula does not apply to a light source that radiates different fluxes in different directions. In such a situation, the concept of *solid angle* is needed. A solid angle is the counterpart in three dimensions of an ordinary angle in two dimensions. The solid angle Ω (Greek capital letter *omega*) subtended by area A on the surface of a sphere of radius r is given by

$$\Omega = \frac{A}{r^2}$$

Solid angle $= \dfrac{\text{area on surface of sphere}}{(\text{radius of sphere})^2}$

The unit of solid angle is the *steradian* (sr); see Figure 17-2. Like the degree and the radian, the steradian is a dimensionless ratio that disappears in calculations.

The general definition of luminous flux is

$$F = I\Omega$$

Luminous flux $= (\text{luminous intensity})(\text{solid angle})$

Since the total area of a sphere is $4\pi r^2$, the total solid angle it subtends is $(4\pi r^2/r^2)$ sr $= 4\pi$ sr. This definition of F thus gives $F = 4\pi I$ for the total flux emitted by an isotropic source.

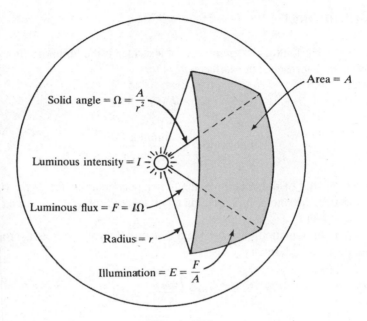

Figure 17-2

You Need to Know ✔

The luminous flux of a 1-cd source gives off per steradian therefore equals 1 lm, and 1 cd equals 1 lm/sr.

The *luminous efficiency* of a light source is the amount of luminous flux it radiates per watt of input power. The luminous efficiency of ordinary tungsten-filament lamps increases with their power, because the higher the power of such a lamp, the greater its temperature and the more of its radiation is in the visible part of the spectrum. The efficiencies of such lamps range from about 8 lm/W for a 10-W lamp to 22 lm/W for a 1000-W lamp. Fluorescent lamps have efficiencies from 40 to 75 lm/W.

Illumination

The *illumination* (or *illuminance*) E of a surface is the luminous flux per unit area that reaches the surface:

$$E = \frac{F}{A}$$

$$\text{Illumination} = \frac{\text{luminous flux}}{\text{area}}$$

In SI, the unit of illumination is the lumen per square meter, or *lux* (lx); in the British system, it is the lumen per square foot, or *footcandle* (fc) (Table 17-1).

The illumination on a surface a distance R away from an isotropic source of light of intensity I is

$$E = \frac{I\cos\theta}{R^2}$$

where θ is the angle between the direction of the light and the normal to the surface (Figure 17-3). Thus, the illumination from such a source varies inversely as R^2, just as in the case of sound waves; doubling the distance means, reducing the illumination to ¼ its former value. For light perpendicularly incident on a surface, $\theta = 0$ and $\cos\theta = 1$, so in this situation, $E = I/R^2$.

Quantity	Symbol	Meaning	Formula	Unit
Luminous intensity	I	Brightness of light source		Candela (cd)
Solid angle	Ω	Three-dimensional equivalent of a plane angle	$\Omega = A/r^2$	Steradian (sr)
Luminous flux	F	Amount of visible light	$F = I\Omega$	Lumen (lm)
Luminous efficiency		Ratio of luminous flux to input power of light source	F/P	Lumen/watt (lm/W)
Illumination	E	Luminous flux per unit area	$E = F/A$	Lux (lx) (= lm/m²) Footcandle (fc) (= lm/ft²)

Table 17.1

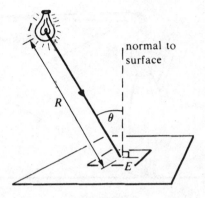

Figure 17-3

Solved Problem 17.1 A 10-W fluorescent lamp has a luminous intensity of 35 cd. Find (*a*) the luminous flux it emits and (*b*) its luminous efficiency.

Solution. (*a*) $F = 4\pi I = (4\pi)(35 \text{ cd}) = 440 \text{ lm}$

(*b*) Luminous efficiency $= \dfrac{F}{P} = \dfrac{440 \text{ lm}}{10 \text{ W}} = 44 \text{ lm / W}$

Reflection of Light

When a beam of light is reflected from a smooth, plane surface, the angle of reflection equals the angle of incidence (Figure 17-4). The image of an object in a plane mirror has the same size and shape as the object but with left and right reversed; the image is the same distance behind the mirror as the object is in front of it.

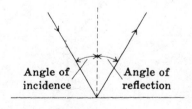

Figure 17-4

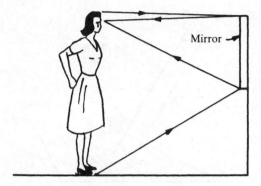

Figure 17-5

Solved Problem 17.2 A woman 170 cm tall wishes to buy a mirror in which she can see herself at full length. What is the minimum height of such a mirror? How far from the mirror should she stand?

Solution. Since the angle of reflection equals the angle of incidence, the mirror should be half her height (85 cm) and placed so its top is level with the middle of her forehead (Figure 17-5). The distance between the mirror and the woman does not matter.

Refraction of Light

When a beam of light passes obliquely from one medium to another in which its velocity is different, its direction changes (Figure 17-6). The

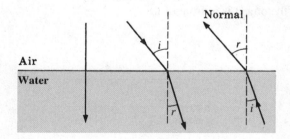

Figure 17-6

greater the ratio between the two velocities, the greater the deflection. If the light goes from the medium of high velocity to the one of low velocity, it is bent toward the normal to the surface; if the light goes the other way, it is bent away from the normal. Light moving along the normal is not deflected.

The *index of refraction* of a transparent medium is the ratio between the velocity of light in free space c and its velocity in the medium v:

$$\text{Index of refraction} = n = \frac{c}{v}$$

The greater its index of refraction, the more a beam of light is deflected on entering a medium from air. The index of refraction of air is about 1.0003, so for most purposes, it can be considered equal to 1.

According to *Snell's law*, the angles of incidence i and refraction r shown in Figure 17-6 are related by the formula

$$\frac{\sin i}{\sin r} = \frac{v_1}{v_2} = \frac{n_2}{n_1}$$

where v_1 and n_1 are, respectively, the velocity of light and index of refraction of the first medium and v_2 and n_2 are the corresponding quantities in the second medium. Snell's law is often written

$$n_1 \sin i = n_2 \sin r$$

In general, the index of refraction of a medium increases with increasing frequency of the light. For this reason, a beam of white light is separated into its component frequencies, each of which produces the sensation of a particular color, when it passes through an object whose sides are not parallel, for instance, a glass prism. The resulting band of color is called a *spectrum*.

Solved Problem 17.3 Why is a beam of white light that passes perpendicularly through a flat pane of glass not dispersed into a spectrum?

Solution. Light incident perpendicular to a surface is not deflected, so light of the various frequencies in white light stays together despite the different velocities in the glass.

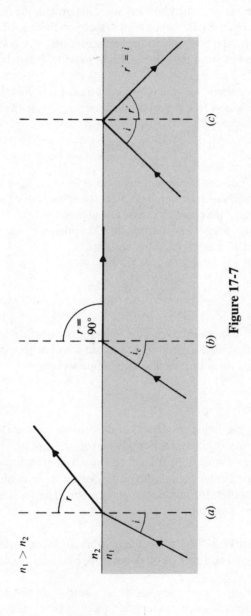

Figure 17-7

Total Internal Reflection

The phenomenon of *total internal reflection* can occur when light goes from a medium of high index of refraction to one of low index of refraction, for example, from glass or water to air. The angle of refraction in this situation is greater than the angle of incidence, and a light ray is bent away from the normal, as in Figure 17-7(a), at the interface between the two media. At the *critical angle* of incidence, the angle of refraction is 90° (Figure 17-7(b)), and at angles of incidence greater than this, the refracted rays are reflected back into the original medium (Figure 17-7(c)). If the critical angle is i_c,

$$\sin i_c = \frac{n_2}{n_1}$$

Apparent Depth

An object submerged in water or other transparent liquid appears closer to the surface than it actually is. As Figure 17-8 shows, light leaving the object is bent away from the normal to the water-air surface as it leaves the water. Since an observer interprets what she or he sees in terms of the straight-line propagation of light, the object seems at a shallower depth than its true one. The ratio between apparent and true depths is

$$\frac{\text{Apparent depth}}{\text{True depth}} = \frac{h'}{h} = \frac{n_2}{n_1}$$

where n_1 is the index of refraction of the liquid and n_2 is the index of refraction of air.

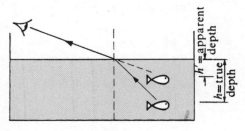

Figure 17-8

Chapter 18
SPHERICAL MIRRORS

In This Chapter:

- ✔ Focal Length
- ✔ Ray Tracing
- ✔ Mirror Equation
- ✔ Magnification

Focal Length

Figure 18-1 shows how a concave mirror converges a parallel beam of light to a real focal point F, and Figure 18-2 shows how a convex mirror diverges a parallel beam of light so that the reflected rays appear to come from a virtual focal point F behind the mirror. In either case, if the radius of curvature of the mirror is R, the *focal length* f is $R/2$. For a concave mirror, f is positive, and for a convex mirror, f is negative. Thus

$$\text{Concave mirror:} \quad f = +\frac{R}{2}$$

$$\text{Convex mirror:} \quad f = -\frac{R}{2}$$

The axis of a mirror of either kind is the straight line that passes through C and F.

C = center of curvature
F = real focal point

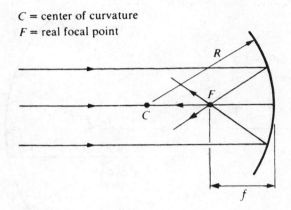

Figure 18-1

C = center of curvature
F = virtual focal point

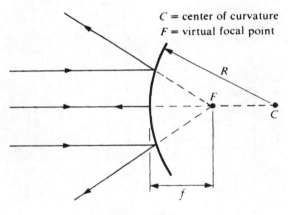

Figure 18-2

Ray Tracing

The position and size of the image formed by a spherical mirror of an object in front of it can be found by constructing a scale drawing by tracing two different light rays from each point of interest in the object to where they (or their extensions, in the case of a virtual image) intersect after being reflected by the mirror. Three rays especially useful for this purpose are shown in Figure 18-3; any two are sufficient:

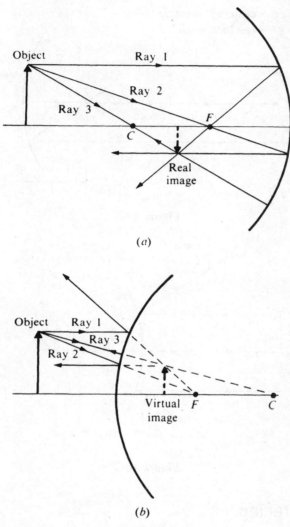

(a)

(b)

Figure 18-3

1. A ray that leaves the object parallel to the axis of the mirror. After reflection, this ray passes through the focal point of a concave mirror or seems to come from the focal point of a convex mirror.

2. A ray that passes through the focal point of a concave mirror or is directed toward the focal point of a convex mirror. After reflection, this ray travels parallel to the axis of the mirror.

3. A ray that leaves the object along a radius of the mirror. After reflection, this ray returns along the same radius.

Mirror Equation

When an object is a distance p from a mirror of focal length f, the image is located a distance q from the mirror, where

$$\frac{1}{p} + \frac{1}{q} = \frac{1}{f}$$

$$\frac{1}{\text{Object distance}} + \frac{1}{\text{Image distance}} = \frac{1}{\text{Focal length}}$$

This equation holds for both concave and convex mirrors (see Figure 18-4). The mirror equation is readily solved for p, q, or f:

$$p = \frac{qf}{q-f} \qquad q = \frac{pf}{p-f} \qquad f = \frac{pq}{p+q}$$

A positive value of p or q denotes a real object or image, and a negative value denotes a virtual object or image. A *real object* is in front of a mirror; a *virtual object* appears to be located behind the mirror and must itself be an image produced by another mirror or lens. A *real image* is formed by light rays that actually pass through the image, so a real image will appear on the screen placed at the position of the image. But a *virtual image* can be seen only by the eye since the light rays that appear to come from the image actually do not pass through it.

Remember

Real images are located in front of a mirror, virtual images behind it.

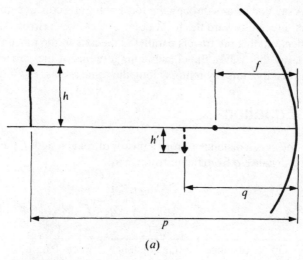

(a)

f = focal length
p = object distance
q = image distance

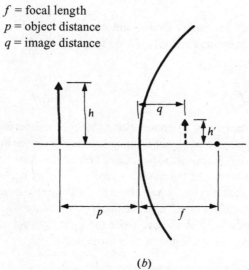

(b)

Figure 18.4

Magnification

The *linear magnification m* of any optical system is the ratio between the size (height or width or other transverse linear dimensions) of the image and the size of the object. In the case of a mirror,

$$m = \frac{h'}{h} = -\frac{q}{p}$$

$$\text{Linear magnification} = \frac{\text{image height}}{\text{object height}} = -\frac{\text{image distance}}{\text{object distance}}$$

A positive magnification signifies an erect image, as in Figure 18-4(*b*); a negative one signifies an inverted image, as in Figure 18-4(*a*). Table 18.1 is a summary of the sign conventions used in connection with spherical mirrors.

Quantity	Positive	Negative
Focal length f	Concave mirror	Convex mirror
Object distance p	Real object	Virtual object
Image distance q	Real image	Virtual image
Magnification m	Erect image	Inverted image

Table 18.1

Chapter 19
LENSES

IN THIS CHAPTER:

- ✔ *Focal Length*
- ✔ *Ray Tracing*
- ✔ *Lens Equation*
- ✔ *Magnification*
- ✔ *Lens Systems*

Focal Length

Figure 19-1 shows how a converging lens brings a parallel beam of light to a real focal point F, and Figure 19-2 shows how a diverging lens spreads out a parallel beam of light so that the refracted rays appear to come from a virtual focal point F. In this chapter, we consider only *thin lenses*, whose thickness can be neglected as far as optical effects are concerned. The focal length f of a thin lens is given by the *lensmaker's equation*:

$$\frac{1}{f} = (n-1)\left(\frac{1}{R_1} + \frac{1}{R_2}\right)$$

In this equation, n is the index of refraction of the lens material relative to the medium it is in, and R_1 and R_2 are the radii of curvature of the two

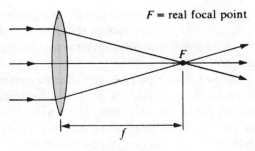

Figure 19-1

surfaces of the lens. Both R_1 and R_2 are considered as plus for a convex (curved outward) surface and as minus for a concave (curved inward) surface; obviously it does not matter which surface is labeled as 1 and which as 2.

A positive focal length corresponds to a converging lens and a negative focal length to a diverging lens.

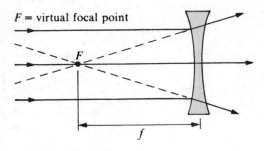

Figure 19-2

Ray Tracing

As with a spherical mirror, the position and size of the image of an object formed by a lens can be found by constructing a scale drawing by tracing two different light rays from a point of interest in the object to where they (or their extensions, in the case of a virtual image) intersect after being refracted by the lens. Three rays especially useful for this purpose are shown in Figure 19-3; any two are sufficient:

1. A ray that leaves the object parallel to the axis of the lens. After refraction, this ray passes through the far focal point of a converging lens or seems to come from the near focal point of a diverging lens.
2. A ray that passes through the focal point of a converging lens or is directed toward the far focal point of a diverging lens. After refraction, this ray travels parallel to this axis of the lens.
3. A ray that leaves the object and proceeds toward the center of the lens. This ray is not deviated by refraction.

Solved Problem 19.1 What is the nature of the image of a real object formed by a diverging lens?

Solution. It is virtual, erect, and smaller than the object, as in Figure 19-3(*b*).

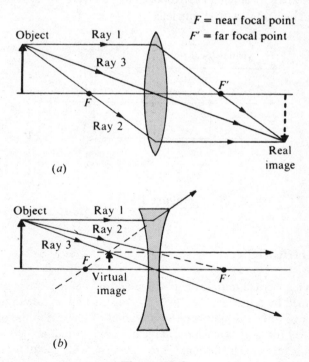

Figure 19-3

Lens Equation

The object distance p, image distance q, and focal length f of a lens (Figure 19-4) are related by the lens equation:

$$\frac{1}{p} + \frac{1}{q} = \frac{1}{f}$$

$$\frac{1}{\text{Object distance}} + \frac{1}{\text{Image distance}} = \frac{1}{\text{Focal length}}$$

This equation holds for both converging and diverging lenses. The lens equation is readily solved for p, q, or f:

$$p = \frac{qf}{q-f} \qquad q = \frac{pf}{p-f} \qquad f = \frac{pq}{p+q}$$

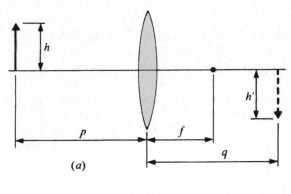

(a)

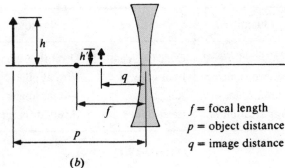

(b)

f = focal length
p = object distance
q = image distance

Figure 19-4

As in the case of mirrors, a positive value of p or q denotes a real object or image, and a negative value denotes a virtual object or image. A real image of a real object is always on the opposite side of the lens from the object, and a virtual image is on the same side. Thus, if a real object is on the left of a lens, a positive image distance q signifies a real image to the right of the lens, whereas a negative image distance q denotes a virtual image to the left of the lens.

Magnification

The linear magnification m produced by a lens is given by the same formula that applies for mirrors:

$$m = \frac{h'}{h} = -\frac{q}{p}$$

$$\text{Linear magnification} = \frac{\text{image height}}{\text{object height}} = -\frac{\text{image distance}}{\text{object distance}}$$

Again, a positive magnification signifies an erect image, a negative one signifies an inverted image. Table 19.1 is a summary of the sign conventions used in connection with lenses.

Solved Problem 18.1 A coin 3 cm in diameter is placed 24 cm from a converging lens whose focal length is 16 cm. Find the location, size, and nature of the image.

Quantity	Positive	Negative
Focal length f	Concave lens	Diverging lens
Object distance p	Real object	Virtual object
Image distance q	Real image	Virtual image
Magnification m	Erect image	Inverted image

Table 19.1

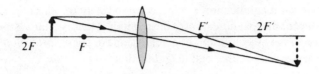

Figure 19-5

Solution. Here $p = 24$ cm and $f = +16$ cm, so the image distance is

$$q = \frac{pf}{p-f} = \frac{(24 \text{ cm})(16 \text{ cm})}{24 \text{ cm} - 16 \text{ cm}} = 48 \text{ cm}$$

The image is real since q is positive (Figure 19-5). The diameter of the coin's image is,

$$h' = -h\frac{q}{p} = -(3 \text{ cm})\left(\frac{48 \text{ cm}}{24 \text{ cm}}\right) = -6 \text{ cm}$$

The image is inverted and twice as large as the object.

In general, an object that is a distance between f and $2f$ from a converging lens has a real, inverted image that is larger than the object.

Lens Systems

When a system of lenses is used to produce an image of an object, for instance, in a telescope or microscope, the procedure for finding the position and nature of the final image is to let the image formed by each lens in turn be the object for the next lens in the system. Thus, to find the image produced by a system of two lenses, the first step is to determine the image formed by the lens nearest the object. This image then serves as the object for the second lens, with the usual sign convention: If the image is on the front side of the second lens, the object distance is considered positive, whereas if the image is on the back side, the object distance is considered negative.

The total magnification produced by a system of lenses is equal to the product of the magnification of the individual lenses. Thus, if the magnification of the objective lens of a microscope or telescope is m_1 and that of the eyepiece is m_2, the total magnification is $m = m_1 m_2$.

PHYSICAL AND QUANTUM OPTICS

Interference

In examining the reflection and refraction of light, it is sufficient to consider light as though it consisted of rays that travel in straight lines in a uniform medium. The study of such phenomena, therefore, is called *geometrical optics*. Other phenomena, notably interference, diffraction, and polarization, can be understood only in terms of the wave nature of light, and the study of these phenomena is called *physical optics*.

    Interference occurs when waves of the same nature from different sources meet at the same place. In *constructive interference*, the waves are in phase ("in step") and reinforce each other; in *destructive interference*, the waves are out of phase and partially or completely cancel (Figure 20-1). All types of waves exhibit interference under appropriate cir-

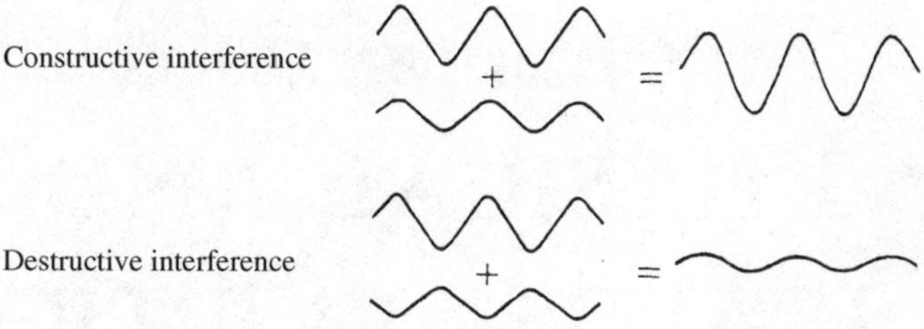

Constructive interference

Destructive interference

Figure 20-1

cumstances. Thus, water waves interfere to produce the irregular surface of the sea, sound waves close in frequency interfere to produce beats, and light waves interfere to produce the fringes seen around the images formed by optical instruments and the bright colors of soap bubbles and thin films of oil on water.

Solved Problem 20.1 When is it appropriate to think of light as consisting of waves and when as consisting of rays?

Solution. When paths or path differences are involved whose lengths are comparable with the wavelengths found in light, the wave nature is significant and must be taken into account. Thus, diffraction and interference can be understood only on a wave basis. When paths are involved that are many wavelengths long and neither diffraction not interference occurs, as in reflection and refraction, it is more convenient to consider light as consisting of rays.

Diffraction

The ability of a wave to bend around the edge of an obstacle is called *diffraction*. Owing to the combined effects of diffraction and interference, the image of a point source of light is always a small disk with bright and dark fringes around it. The smaller the lens or mirror used to form the image, the larger the disk. The angular width in radians of the image disk of a point source is about

$$\theta_o = (1.22)\left(\frac{\lambda}{D}\right)$$

where λ is the wavelength of the light and D is the lens or mirror diameter. The images of objects closer than θ_0 will overlap and hence cannot be resolved no matter how great the magnification produced by the lens or mirror. In the case of a telescope or microscope, D refers to the diameter of the objective lens. If two objects d_0 apart that can just be resolved at a distance L from the observer, the angle in radians between them is $\theta_0 = d_0/L$, so the above formula can be rewritten in the form

$$\text{Resolving power} = d_o = (1.22)\left(\frac{\lambda L}{D}\right)$$

Polarization

A *polarized* beam of light is one in which the electric fields of the waves are all in same direction. If the electric fields are in random directions (though, of course, always in a plane perpendicular to the direction of propagation), the beam is *unpolarized*.

You Need to Know

Various substances affect differently light with different directions of polarization, and these substances can be used to prepare devices that permit only light polarized in a certain direction to pass through them.

Quantum Theory of Light

Certain features of the behavior of light can be explained only on the basis that light consists of individual *quanta*, or *photons*. The energy of a photon of light whose frequency is f is

$$\text{Quantum energy} = E = hf$$

where h is *Planck's constant*:

Planck's constant $= h = 6.63 \times 10^{-34}$ J·s

A photon has most of the properties associated with particles—it is localized in space and possesses energy and momentum—but it has no mass. Photons travel with the velocity of light.

The electromagnetic and quantum theories of light complement each other: Under some circumstances, light exhibits a wave character, under other circumstances, it exhibits a particle character. Both are aspects of the same basic phenomenon.

X-Rays

X-rays are high frequency electromagnetic waves produced when fast electrons impinge on a target. If the electrons are accelerated through a potential difference of V, each electron has the energy KE $= eV$. If all this energy goes into creating an X-ray photon, then

$$eV = hf$$
Electron kinetic energy = X-ray photon energy

and the frequency of the X-rays is $f = eV/h$.

Index